Storm Watchers

The Turbulent History of Weather Prediction from Franklin's Kite to El Niño

•

JOHN D. COX

John Wiley & Sons, Inc.

To my mother and father,

ELIZABETH COX and ERNEST Y. COX

Published by John Wiley & Sons, Inc., Hoboken, New Jersey
Published simultaneously in Canada

For general information about our other products and services, please contact our Customer Care Department within the United States at (800) 762-2974, outside the United States at (317) 572-3993 or fax (317) 572-4002.

Wiley also publishes its books in a variety of electronic formats. Some content that appears in print may not be available in electronic books.

Library of Congress Cataloging-in-Publication Data:

Cox, John, date.
Storm watchers : the turbulent history of weather prediction from Franklin's kite to El Niño / John Cox.
p. cm.
Includes index.
ISBN 0-471-38108-X (hardcover)
1. Weather forecasting—History. I. Title.
QC995 .C79 2002
551.63'09—dc21

2002005222

Printed in the United States of America

10 9 8 7 6 5 4 3 2 1

Contents

•

PART IV
Together at the Front

PART V
Suddenly New Science

Introduction

•

WEATHER FORECASTING has become a kind of appliance science, part of the electric rhythm of life, absorbed and applied without a second thought to the mundane questions of personal comfort and convenience. How cold will it be today? Or tomorrow? Or how hot? Will I need a sweater? An umbrella? A hat?

Depending on the medium, its presentation can be brief and stylized to the point of wordless icons, or extended and elaborate, colorfully portrayed with animated and entertaining visual effects. These productions relate knowledge that is different from the rest of the daily information stream. This is the future, the telling of atmospheric motions and events that have not happened yet.

Weather forecasting is the product of meteorological science that comes from the edge of our most advanced capability, quantifying circumstances of turbulence and chaos that are at the limits of probability. It is the numerical modeling output of some of the most powerful computers on the planet. This knowledge of the future is costly and difficult to acquire. It is so costly, in fact, its raw material so extensive, its reach so global, and its processing so computationally demanding that only governments can provide it.

Day in and day out, weather prediction is more accurate and more useful than ever, more objective and more thoroughly scientific, although this trend of continuing improvement is not commonly recognized. So generally reliable is this prediction of the future, in fact, that its accuracy is taken for granted, like a public utility or a civil right. When it proves to be inaccurate, the occasion is widely remarked upon and not quickly forgotten.

The daily weather forecast is a marvel of digital electronics, a set of facts that is intensely modified by computer from the first assimilation of raw data to its final graphical output on a screen. Routinely employed are data from sophisticated instruments aboard satellites, automated weather stations, radio-equipped balloons, airliners, ocean buoys, and ships at sea. Its advanced and most accurate form, Numerical Weather Prediction, would not even be possible without the incredible speed and sophistication of supercomputers.

Notwithstanding all of this artificial intelligence, the science of forecasting weather is a triumph of human intelligence—the work of people. While modern weather science is the work of many, it is founded on the work of a relative few. These are the stories of some of the men who discovered how the atmosphere works and how to foretell its behavior. Their science has saved and continues to save many lives. An ancient dream of accurately predicting weather has been fulfilled. The storm watchers deserve to be remembered, many of them as heroes.

PART I

A Newborn Babe

•

Knowledge of the weather is very old information on the human scale, and fear of it is older still. Was there ever a time when humans were not thinking about it one way or another, offering sacrifice or prayer, or passing around advice? Scientific study of the weather, however, is a fairly recent turn of events, hardly more than two centuries old. The term *meteorology* goes all the way back to Aristotle's *Meteorologica,* written about 340 B.C., meaning the study of things that fall from the sky, although really nothing of the science comes down from that treatise.

Weather science is young because it is so demanding. For most of history, motions of the atmosphere defied accurate description. Although the primary instruments for measuring atmospheric changes—the thermometer, the barometer, and the hygrometer—go back 400 years and more to Galileo, Torricelli, and other European savants, understanding the usefulness of these instruments was a long time coming. Meteorology had to wait for other sciences to define its basic principles and forces: relations of energy and mass, gravity and friction, thermodynamics, and the behavior of gases. At every step along the way, it seems, students of the atmosphere underestimated the difficulty of the problems they were trying to solve, the depth of their mysteries. Progress has been irregular and frequently disappointing. Cobbling this principle to that law borrowed from physics or astronomy or chemistry, people who studied weather were generally taking time out from more reliably fruitful pursuits.

Many years would pass before weather gained sustained attention from scientists. In its early stages from the middle of the eighteenth century to the middle of the nineteenth century, the study of weather was the avocation of learned gentlemen and natural philosophers, people with secure reputations in other fields. Every advance carried the burden of displacing other, more venerable explanations of the ways of the winds and the causes of storms. Doubters outnumbered supporters. The early weathermen were individuals who saw things differently than most people of their time. What they had in common was a special talent for observation and an intuitive sense that discerned patterns and processes where others saw only randomness and tumult. When the science was young, the study of weather was an act of sturdy independence and courage.

1

Benjamin Franklin

Chasing the Wind

•

IT IS TYPICAL of the history of meteorology that the modern study of storms should begin with the description of a spoiled astronomical event.

The study of weather has always been measured, invariably to its detriment, by the standards of astronomy, its older and more respected sister science. Through thousands of years of kingdoms advised by astronomers, there was never a Meteorologist Royal. Knowledge of the heavens was far advanced by the time the investigation of weather was deemed worthy of a serious man's preoccupation. In the middle of the eighteenth century, astronomy was preeminent and meteorology was hardly a science at all. Some things were respectably knowable about the physical world and some were not.

Two hundred fifty years ago, astronomers could predict the occasions of lunar eclipses precisely as to date and time of day, and they could explain their cause and effects: that the moon's orbit passes periodically out of the brightness of the sun's light and into the darkness of the celestial shadow cast by Earth. By such divinely predictable events the clockwork universe was affirmed. About the intervening weather, on the other hand, that a storm might blow up and obscure an eclipse—whether, when, where, or why—no one had any idea about such unaccountable acts of God.

In the autumn of 1743, as the moon approached Earth's shadow, Benjamin Franklin, a 37-year-old printer and newspaper publisher in the American colony of Pennsylvania, was making plans to witness the eclipse from his home in Philadelphia. A busy and prosperous man, he was nonetheless an eager observer of such phenomena. This eclipse would begin about 8:30 P.M. on October 21, a Friday night. But as Franklin recalled a few years later in a letter to his friend Jared Eliot, "before night a storm blew up at northeast, and continued violent all night and all the next day; the sky thick-clouded, dark and rainy, so that neither moon nor stars could be seen."

Obscured as it was, the timing of the eclipse illuminated for the perceptive Franklin something entirely unexpected about the violent storm, its whereabouts, and its movement that night. As he wrote to Eliot:

> The storm did great damage all along the coast, for we had accounts of it in the newspapers from Boston, Newport, New York, Maryland and Virginia; but what surprised me was, to find in the Boston newspapers, an account of the observation of that eclipse made there; for I thought as the storm came from the northeast, it must have begun sooner at Boston than with us, and consequently have prevented such an observation. I wrote my brother (in Boston) about it, and he informed me that the eclipse was over there an hour before the storm began.

From this information, a scrap of detail that might have gone unnoticed in the hands of a less vitally interested observer, grew an idea that would be central to meteorology, especially to weather forecasting. Storms have characteristic structures and preferred directions of travel. Franklin formed a generalization about the movement of what he continued to describe to Eliot as "storms from the northeast" that could blow violently, sometimes for three or four days. Franklin wrote: "Of these I have had a very singular opinion for some years, viz: that, though the course of the wind is from northeast to southwest, yet the course of the storm is from southwest to northeast; the air is in violent motion in Virginia before it moves in Connecticut, and in Connecticut before it moves at Cape Sable, etc." More than 150 years later, in 1899, the Harvard scholar William Morris Davis, writing in the *Journal of the Franklin Institute,* would look back on this suggestion as a defining moment, observing that "with this began the science of weather prediction." In Franklin's day, however, weather prediction was beyond the realm of science.

As a practical matter, the storm traveled faster than could words of warning in the eighteenth century, although clearly not as fast as Franklin's estimate of 100 miles an hour. The great man was not so great with numbers. In any case, obstacles to scientific weather prediction were more deeply rooted in both the Old World and the New. Future weather was treated like the future of anything else, part of the occult prognostications of astrologers, especially in Europe, where their profitably published almanacs offered artfully worded weather predictions for the entire year. In the colonies, Franklin himself enjoyed a handsome income for 25 years as publisher of *Poor Richard's Almanack,* although his prognostications of weather always came with characteristic humor and wit. Praising Franklin's contributions to meteorology, the pioneering American weather scientist Cleveland Abbe took a close look at the Franklin almanacs in 1906 and found no astrology in them. In a presentation to the American Philosophical Society of Philadelphia, Abbe said, "Now while it is true that in these he published conjectures as to the weather during the respective years, yet we are not to think of Franklin as a planetary meteorologist, for the fact is that in every one of these issues he disclaims all knowledge of the weather or astrology and pokes fun at his own predictions as utterly absurd and useless."

Explanations for the causes of weather remained a traditional part of church doctrine, as it had through the Middle Ages. Since its rediscovery in the twelfth century, Aristotle's *Meteorologica* had been installed as Christian dogma, and his conjectures about the organismic exhalations of Earth satisfied nearly 700 years of theological meteorology. (As a meteorologist, Aristotle was a pretty good philosopher. Not even the loyal pupil Theophrastus could accept his mentor's bald assertion that the wind was not moving air.) In the American colonies, the Puritan clergy yielded to no "secondary natural causes" the power and word of the Almighty in the fierce tempests of the New World, even under the most terrible circumstances.

On August 31, 1735, in New London, Connecticut, a great storm rose overhead just as the Reverend Eliphalet Adams was beginning his afternoon service. A bolt of lightning, "the fire of God," shot down upon his church. Timbers crashed down among the congregation. Smoke and dust filled the air. So fierce was the thunderclap that it left their ears ringing. Everywhere in the wrecked room were wounded of his flock, burned and broken. Pitiful shrieks of shock and agony rang out. At his feet, at the very horn of his altar, a young man, Edwin Burch, lay dying.

The following Sabbath, those of the congregation who were not too seriously injured came back to the temporarily patched meetinghouse to hear about the god who strikes down the faithful at prayer. A proud Adams, with a view to posterity, saw to the printing of his sermon "God Sometimes Answers His People by Terrible Things in Righteousness. A Discourse Occasioned by that Awful Thunderclap which Struck the Meeting-house in N. London, August 31st, 1735." Adams saw God's righteous hand in the sheer power of the event and his merciful hand in the fact that the calamity was not worse: "We might have died by scores and by hundreds, yea, the whole congregation might have been dispatched at once into eternity."

For the disaster, and for the death of Edwin Burch, the Puritan faithful had only their unworthy selves to blame. Adams sermonized:

> There is no blemish or defect in any thing which God doth, nothing of which we should dare to say, that it ought to have been otherwise; there are faults enough and enough in us to justify the Lord in his most severe dispensations towards us; we must hold our peace and not open our mouths to complain, nor suffer an unease or grudging thought to stir in our hearts, how heavy so ever the strokes be, or how much so ever we are made to smart thereby, we must still ascribe righteousness to our maker and our judge.

Just 10 years later, in nearby Philadelphia, Benjamin Franklin would begin a series of investigations into the nature of electricity, work that made him famous. Before long, in 1749, he was entering into his notebook certain conjectures about its similarities to lightning: "The electric fluid is attracted by points; we do not know whether this property is in lightning; but since they agree in all the particulars wherein we can already compare them, is it not probable that they agree likewise in this? Let the experiment be made." In June, 1752, he performed his famous kite experiment, causing an electrostatic spark between a knuckle and a key hanging from the twine. This and other experiments of his design soon established lightning as an electrical phenomenon in the atmosphere. The discovery would lead to the installation of insulated and grounded iron "points," or lightning rods, that carried the fire of God harmlessly down the sides of vulnerable church steeples across the land.

Benjamin Franklin's seven years of research into the nature of electricity, his most intensely focused period of scientific activity, earned him an international reputation as a scientist. In later years, he stopped practicing science only in the sense that in leading the American colonies through revolution and into nationhood, he lost his time to pursue it. All his life, he remained one of the most observant students of nature. Nothing interesting about the weather, or about much of anything else, seems ever to have escaped his curiosity. He thought and wrote about weather for 60 years. In 1726, on a return voyage from London to Philadelphia, entries in his journal included routine weather observations and the appearance of an unusual "lunar rainbow." In 1786, he was offering long-range forecasts to members of his family.

In the spring of 1755, he and a group of friends who were riding on the Maryland country estate of Colonel Benjamin Tasker watched a whirlwind approaching, growing as it came toward them up a hill. Franklin described this adventure in a letter in August to a friend, Peter Collinson.

> The rest of the company stood looking after it; but my curiousity being stronger, I followed it, riding close by its side, and observed its licking up in its progress all the dust that was under its smaller part. As it is a common opinion that a shot, fired through a water-spout will break it, I tried to break this little whirlwind by striking my whip frequently through it, but without any effect. Soon after, it quitted the road and took into the woods, growing every moment larger and stronger, raising instead of dust the old dry leaves with which the ground was thick covered, and making a great noise with them and the branches of the trees, bending some tall trees round in a circle swiftly and very surprisingly, though the progressive motion of the whirl was not so swift but that a man on foot might have kept pace with it; but the circular motion was amazingly rapid. By the leaves it was now filled with I could plainly perceive that the current of air they were driven by moved upwards in a spiral line; and when I saw the trunks and bodies of large trees enveloped in the passing whirl, which continued entire after it had left them, I no longer wondered that my whip had no effect on it in its smaller state.

Franklin rejoined the company, and as the group traveled on for nearly three miles, he watched the leaves taken up by the whirlwind continue to fall from the sky. He wrote Collinson: "Upon my asking Colonel Tasker if such whirlwinds were common in Maryland, he answered pleasantly: 'No, not at all common; but we got this on purpose to treat Mr. Franklin.' And a very high treat it was too."

Accounts of whirlwinds, tornadoes, and waterspouts appear in the chronicles of the ancient world, although nowhere were they as common as in North America, where, since the founding of the colonies, they had been subjects of speculation. In the middle of the eighteenth century, Franklin was in the thick of it, providing an early description of the process of convection. In a letter written in 1753, he proposed two sets of conditions for such winds:

> 1. That the lower region of air is often more heated, and so more rarified, than the upper; consequently, specifically lighter. The coldness of the upper region is manifested by the hail, which sometimes falls from it in a hot day. 2. That heated air may be very moist, and yet the mois-

> ture so equally diffused and rarified, as not to be visible till colder air mixes with it, when it condenses and becomes visible. Thus our breath, invisible in summer, becomes visible in winter.

Although he was always a close observer, Franklin was still a natural philosopher at heart, and he was not inclined to clutter his conjectures with a lot of data or mathematics. He knew good science when he saw it, and he knew when his own theorizing wandered beyond observed facts into "the wilds of fancy." But the man who completed only two years of formal education never lost his disdain for mathematics, the *lingua franca* of modern meteorology. In his letter describing his thinking about whirlwinds, he concluded, "If my hypothesis is not the truth itself it is least as naked: For I have not, with some of our learned moderns, disguised my nonsense in Greek, clothed it in algebra, or adorned it with fluxions. You have it in *puris naturalibus*."

Franklin also found himself attracted to questions of climate, the longer-term state of weather, a subject which had bedeviled the colonists and their European sponsors since the founding of the settlements in the sixteenth century. Without a grasp of the general circulation of the atmosphere, the west-to-east flow in the middle latitudes, nothing was known of the more extreme continental character of the New World's climate. Basking in their moderate oceanic climes, warmed by the Gulf Stream, Europeans were confounded by the patterns of seasonal weather in the colonies. By Franklin's day, however, a general warming trend was noticeable, and in 1763, he met with a group of colonial scholars to discuss the changing climate. Franklin agreed with others that deforestation was likely the cause, that "cleared land absorbs more heat and melts snow quicker," although he argued that many more years of observations would be necessary to make the case.

Other investigations by Franklin at this time led to important advances in the understanding of the Gulf Stream, the "river" of warm water that travels from the tropics far north along the American coast and across the North Atlantic. As deputy postmaster general of the colonies, Franklin heard complaints that English postal vessels traveling from Falmouth to New York consistently took several more days crossing the Atlantic than merchant vessels making the longer voyage from London to Rhode Island. A Franklin acquaintance, Tim Folger, a Nantucket whaler, had a ready explanation. He sketched out the Gulf Stream, and Franklin had it engraved and published on a map that drew wide attention to the important navigational feature from everyone except the captains of British packets, who were not about to take any advice from American fishermen in that day and age. This work led Franklin to take regular readings of sea-surface temperatures on his voyages across the Atlantic, marking the first use of the thermometer as a navigational device.

In his late 70s, while serving as ambassador to France and living near Paris, Franklin noticed a peculiar lack of intensity to the sunlight in the summer of 1783 and drew a connection between it and the severity of the following winter across Europe. He described his thinking in a memoir published by the Manchester Society: "During several of the summer months of the year 1783," he wrote, "when the effects of the sun's rays to heat the earth in these northern regions should have been the greatest, there existed a constant fog over all Europe, and great part of North America. This fog was of a permanent nature; it was dry, and the rays of the sun seemed to have little effect towards dissipating it, as they easily do a moist fog, arising from water." This coolness caused the earth to absorb less heat, he reasoned. "Hence the surface was early frozen. Hence the first snows remained on it unmelted, and received continual additions. Hence perhaps the winter of 1783–4, was more severe than any that happened for many years."

"The cause of this universal fog is not yet ascertained," he wrote. Perhaps it was the burned-out debris of a comet or asteroid, he supposed, or more particularly, "the vast quantity of smoke, long continuing to issue during the summer" from volcanoes near Iceland.

"It seems however worth the inquiry, whether other hard winters, recorded in history, were preceded by similar permanent and widely extended summer fogs," Franklin wrote. "Because, if found to be so, men might from such fogs conjecture the probability of a succeeding hard winter . . . and take such measures as are possible and practicable, to secure themselves and effects from the mischiefs that attend the last."

Following Franklin's line of thought, modern earth and weather scientists, searching climate evidence and historical records, have indeed found a pattern of atmospheric cooling lasting up to two years after the eruptions of large volcanoes.

Benjamin Franklin died on April 17, 1790, at his home in Philadelphia at the age of 84. In 1906, in Philadelphia, the American Philosophical Society, a learned group which Franklin founded in 1743, celebrated the bicentennial of his birth. One of the nation's most astute scientists, Cleveland Abbe, himself a pioneer of American weather forecasting, took the occasion to describe Franklin's contribution to meteorology. "To the laurel that crowns him," Abbe added another leaf: "as the pioneer of the rational long-range forecasters, and of the physical meteorologists who will, undoubtedly, in the future develop this difficult subject."

2

Luke Howard

Naming the Clouds

•

LUKE HOWARD ALWAYS lamented the fact that he had been taught "too much Latin grammar and too little of anything else" at the private Quaker school of his youth in Oxfordshire, England. His real interest was science, although it would never be more than an avocation in his life. In a letter to the great German dramatist and poet Johann Wolfgang von Goethe, he wrote in 1822 that "from the first my real penchant was towards meteorology." Pursuing this lifelong avocation, Howard would make observations that would bring order to the sky like nobody's before or since. Howard would transform the clouds from amorphous, ever-changing masses of vapor into objects of coherent particularity formed by knowable physical processes. It was a signal achievement, crucial to the founding of the new science. And in the winter of 1802, when it came time to describe them, to give to the clouds the names that would distinguish one type from another, he would, ingeniously, fall back on the old language of his classical education. And for two centuries and beyond, every meteorologist in every country of the world would learn to speak Luke Howard's Latin: *cirrus, cumulus, stratus, nimbus.*

Goethe wrote that Luke Howard "was the first to hold fast conceptually the airy and always changing form of clouds, to limit and fasten down the indefinite, the intangible and unattainable and give them appropriate names." Howard's taxonomy of clouds brought to the sky what Carl Linnaeus brought to biology and Charles Lyell to geology: not just a list of names, but an orderly new way of looking at nature. His system of nomenclature and the insights it represented spread fast and far, not only through the infant science of meteorology but beyond into the rich culture of nineteenth-century Europe. Goethe composed four poems about the clouds, dedicating them to Howard. Within a few years, for the first time, accurate and meticulously detailed representations of clouds began appearing in the skies of romantic era paintings of such masters as Caspar David Friedrich in Germany, and Joseph M. W. Turner and John Constable in England. Across the Atlantic, Howard's clouds inspired the skies of the nineteenth-century American landscape masters Thomas Cole, Frederick Church, and George Inness. "The sky too belongs to the Landscape," Howard wrote. "The ocean of air in which we live and move, in which the bolt of heaven is forged, and the fructifying rain condensed, can never be to the zealous Naturalist a subject of tame and unfeeling contemplation."

Born in London in 1772, Luke was the first child of Elizabeth and Robert Howard, a devout Quaker and a successful manufacturer who introduced to England an efficient new oil-burning lamp invented by the Swiss engineer Aimé Argand. While Luke would follow his father into business and become a successful manufacturer of pharmaceutical chemicals, the study of meteorology would always be an important part of his life. In a book about the barometer late in his life, Howard wrote that he was "addicted to this study from my boyhood." And he recalled in his letter to Goethe that as a boy of 11 he was "much interested by the remarkable summer haze and aurora borealis of 1783." As Benjamin Franklin had surmised, volcanic eruptions in Iceland and Japan that year and the next cast a pall of ash over Europe. At the time the dense haze was known as the "Great Fogg."

After completion of his studies in the large Quaker school near Oxford, Luke was apprenticed for seven years to a retail pharmacist in Stockport, near Manchester. He returned to London and in 1796 went into business with William Allen, who operated a pharmacy in the city. Howard took charge of the chemical manufacturing laboratory at Plaistow, an industrial area in Essex, east of London. Allen and Howard also formed a small philosophical group called the Askesian Society to cultivate their mutual scientific interests among like-minded friends. It was before this group, meeting in Allen's home in December 1802, that Howard presented his famous "Essay on the Modification of Clouds."

Howard's meteorological interests went beyond the clouds, although his most important contributions did not extend far beyond the subject. A methodical and uncommonly close observer, he would not be remembered as a great theorist. And he never pretended to be a great scientist, always adopting a tone of modesty in his presentation and pointing to areas of the infant science requiring more research. He was a well-to-do and busy businessman, occupied daily with the affairs of property and commerce. He was a devout Quaker and a prolific writer on religious and social subjects of contemporary importance to the Society of Friends. As he wrote Goethe in 1822, his "claims to be a man of science . . . are merely small, however, since I was born with a capability of observation, I then began to make use of it."

Earlier in 1802 he had presented to the Askesian Society an essay titled "Proximate Causes of Rain, and on Atmospheric Electricity." Howard's ideas about the causes of rain were not persuasive. Like many scientists of the era, Howard was very much impressed with the recent discovery of electricity in the atmosphere. Long after Franklin's experiments, researchers were pursuing the proposition that electricity in the air was a cause rather than a by-product of important weather processes, including the production of rain and the motions of storms. So Howard found that "rain is in almost every instance the result of the electrical action of clouds upon each other," although he would emphasize that it is secondary to the "two grand predisposing causes—a falling temperature and the influx of vapour."

In 1818 and 1819 he would publish the two-volume book *Climate of London,* the first study of its kind. In this extensive report, Howard would break new ground in the investigation of the effects of urban air pollution on weather. He would become the first to identify a phenomenon that later atmospheric scientists would call the *heat island effect,* the ability of industrial and urban materials to radiate heat to such a degree as to cause changes in local weather.

Howard was a pioneer among the many weather scientists over the years who ardently searched for, and almost invariably found, discernible cycles in weather and climate. In 1842 he published a theory based on a series of personal weather observations that claimed to establish an 18-year cycle in the seasons of Britain. It was not a convincing paper. Nor was his research published 2 years earlier that purported to establish a 9-year cycle. Before this kind of statistical chase had run its course, climate and weather cycles of almost every period would be discovered and triumphantly described.

For all of that, few contributions to meteorology have lasted like Howard's classification of clouds. This system is constructed with the care of someone who seemed to be conscious of its permanence. In the eighteenth century achievements of Linnaeus, Howard was aware of a perfect model for his scheme. The great Swedish naturalist Carl von Linné had brought a classic, methodical nomenclature to the plant and animal kingdoms, and his Latin-based taxonomy had swept through the scientific world. In 1800, before his cloud study, Howard had presented his first scientific paper, on the subject of pollens, to the Linnean Society in London.

At the time that Howard was working on clouds in England, the same idea was being developed by Jean-Baptiste Lamarck, the great evolutionary biologist in France, although the work of each was unknown to the other. While the two systems were not very different, Lamarck's work failed to make an impression with fellow scientists, even in France. Like Howard, Larmarck observed that different cloud types form at different heights of the atmosphere, which he divided into three layers, and several of his cloud types are quite similar to Luke Howard's. But where Howard had selected common words in universally acceptable Latin, Larmarck had chosen unusual French names for the types of clouds. While Howard was clear that knowledge of clouds was an important part of meteorological research and theory, Lamarck, in an 1802 paper, "On Cloud Forms," seemed almost defensive on the subject. "It is not in the least amiss for those who are involved in meteorological research to give some attention to the form of clouds," Lamarck wrote, "for, besides the individual and accidental forms of each cloud, it is clear that clouds have certain general forms which are not all dependent on chance but on a state of affairs which it would be useful to recognize and determine." Perhaps Lamarck had good reason to be defensive. He would suffer a reputation in France for "spending much time in fruitless meteorological prediction." Finally, a rude public rebuke by Napoléon, that he should stick to natural history, caused him to abandon his meteorological researches. Howard's essay was heard by fellow Askesian Society member Alexander Tilloch, editor of *Philosophical Magazine,* who soon published the paper in his respected journal. Acceptance of Lamarck's research, in contrast, certainly suffered from the fact that it was published in a journal that also contained treatises on astrological meteorology. Later, other systems of cloud classifications were proposed by such researchers as Heinrich Dove in Germany in 1828 and Elias Loomis in the United States in 1841; but for the better part of the whole science of meteorology, Howard's would be the international standard. Howard's Latin names and the layered scheme he developed, defining clouds by their height and shape, was the basis of the classification system finally adopted in 1929 by the International Meteorological Commission.

"Since the increased attention which has been given to Meteorology, the study of the various appearances of water suspended in the Atmosphere has become an interesting and even necessary branch of that pursuit," Howard began in his seminal paper. Long before the physics of cloud formation was well established, Howard recognized that clouds of different shapes occupy different regions of the sky and form through different processes. That the classifications survive the enormous gaps in understanding of these processes is testimony to the care he always took in his research. Howard wrote that clouds "are subject to certain distinct modifications, produced by the general causes which affect all the variations of the Atmosphere: they are commonly as good visible indications the operation of these causes, as are facial expressions, of the state of a person's mind or body." A meteorologist relying only on his instruments, he wrote, was taking only "the pulse" of the atmosphere.

First among the types came *cirrus,* meaning "lock of hair," clouds that Howard describes as "parallel, flexuous, or diverging fibres, extensible by increase in any or in all directions." Then came *cumulus,* meaning "heap," that Howard called "convex or conical heaps, increasing upward from a horizontal base." And then followed *stratus,* or "layer," "a widely extended, continuous, horizontal sheet, increasing from below upward." *Nimbus,* or "rain," was "the rain cloud. A cloud, or system of clouds from which rain is falling. It is a horizontal sheet, above which the Cirrus spreads, while the Cumulus enters it laterally and from beneath."

"While any of the clouds, except the nimbus, retain their primitive forms, no rain can take place," he wrote, "and it is by observing the changes and transitions of cloud form that weather may be predicted." Howard explained that the names "were intended as arbitrary terms for the structure of clouds, and the meaning of each was carefully fixed by a definition." He hoped they would become part of a "universal language, by means of which the intelligent of every country may convey to each other their ideas without the necessity of translation. And the more this facility of communications can be increased, by our adopting by consent uniform modes, terms, and measures for our observations, the sooner we shall arrive at a knowledge of the phenomena of the atmosphere in all parts of the globe, and carry the science to some degree of perfection."

3

James Glaisher

Taking to the Air

•

ON THE AFTERNOON of September 5, 1862, in the gloriously bright light and brilliantly blue sky high above a sea of clouds over England, a dark and deadly curtain was beginning to descend. Suddenly the scientist was struggling to make his observations. The images of the fine measuring scales of the barometer and the thermometer were blurring and fading in his eyes. James Glaisher was more than five miles in the air and still the balloon was rising. Panting for breath, he called to his pilot to help with the instruments, but through his failing eyes he saw that Henry Coxwell had climbed up onto the ring above the car to work with the rigging. Glaisher's arms went limp. Then his legs splayed out from under him and he fell against the edge of the car. He couldn't speak. Resting on his shoulder, his head drooped out over the abyss.

In putting himself in this predicament, Glaisher's goal was to observe the critical features of the upper air, the temperatures, the pressures, and the dew points up through the realm of the atmosphere that always had been out of the reach of meteorologists. Until a scientist was willing to assume this risk, the closest researchers could come were the observations made occasionally from the tops of high mountains. Finally they had a mode of flight and a mission dedicated to reaching the heights of the free air and making the observations that would give them a real picture of the structure of the atmosphere. Scientific expectations were high.

Although manned balloon flight had been invented in France in 1783, its employment in the service of science had come slowly.

Benjamin Franklin, the 77-year-old American ambassador to France, was present to witness the spectacle that afternoon in August 1783 when the first unmanned hydrogen gas balloon was released over Paris.

"Of what use is such a device?" a skeptic asked.

"Of what use is a newborn babe?" Franklin replied.

The value of such a device to the purposes of meteorology was soon recognized by the few scientists pursuing the subject at the time. The physicist Jacques Charles, who invented the gas balloon, took a thermometer and a barometer on his first ascent in 1783. The American physician John Jeffries took temperature readings and air samples during a 1784 ascent in France. And the chemist Joseph-Louis Gay-Lussac and the physicist Jean-Baptiste Biot made a few scientific flights in 1804. Still, most balloonists were interested in temperature and wind variables of the upper air only to the extent that they affected the progress of their ascents. However valuable the scientific observations might be, most balloons and balloonists were hired for their spectacle value, to entertain crowds at large public events. It was several decades before the newborn babe reached maturity as a scientific tool. Manned balloon ascents were costly and unpredictable, and few philosophers were willing to stake their lives on such a contrivance.

Only in the middle of the nineteenth century was the device put to work for science in a serious way. The British Association for the Advancement of Science had financed a few flights in 1852, when John Welsh, superintendent of its observatory at Kew, obtained records of temperature and humidity up to 20,000 feet. Now the association was financing Glaisher and Coxwell to investigate a long list of questions about the upper atmosphere. On the afternoon of September 5, 1862, however, at 1:57 P.M., one of the first balloon ascents dedicated to meteorology was looking like a terrible mistake.

"In an instant intense darkness overcame me," Glaisher would recall, but still, for a few seconds longer, he was conscious. "I thought I had been seized with asphyxia, and believed I should experience nothing more, as death would come unless we speedily descended." Nine and three-quarters inches of mercury was the last air pressure reading he had been able to see. By Glaisher's estimation they were above 29,000 feet, higher in the sky than anyone had ever been before. Sightless, unable to move or speak, the 53-year-old meteorologist lost consciousness. His body was starving of oxygen, just as he suspected, but the most perilous circumstance was the fact that at that very moment the pilot of the balloon was completely unaware of the desperate condition of his passenger.

Henry Coxwell was up in the rigging of the balloon, sitting on the ring, working on a serious mechanical problem. From the very moment that Glaisher and Coxwell had left the ground near Wolverhampton at 1:03 P.M., the balloon had been rotating, almost spinning like a top. This constant motion had caused the rip cord, the line to the shutters of the gas valve, to become entangled in the rigging. Unless Coxwell could open the gas valve, the balloon would remain completely out of control, spinning and rising ever higher in the silence and the growing cold and the increasingly rarefied air.

Coxwell had taken off a pair of thick gloves in order to better handle the sand bags, "and the moment my unprotected hands rested on the ring, which retained the temperature of the air, I found that they were frostbitten." The neck of the balloon was ringed with frost and the air was piercingly cold. By the time the valve line finally hung free, the pilot had lost all use of his hands. Coxwell placed his arms on the ring and allowed himself to fall down into the wicker car, bringing the line down with him. His breath coming in desperate gulps, he spoke to Glaisher, and only then realized the perilous condition of the scientist.

Now Coxwell felt himself losing consciousness. His hands were numb and black with frostbite, and his arms and legs were becoming useless. With the last of his faculties he seized the valve cord with his teeth and jerked his head down once, twice, and again. Finally the hydrogen began escaping, and the balloon began to descend.

"I then looked round, although it seemed advisable to let off more gas, to see if I could in any way assist Mr. Glaisher, but the table of instruments blocked the way, and I could not, with disabled hands, pass beneath," Coxwell recalled. "My last hope, then, was in seeking the restorative effects of a warmer stratum of atmosphere." He tugged again on the valve line. He felt himself strengthening as the balloon continued falling with a swoosh, and he called out to the scientist.

"Never shall I forget those painful moments of doubt and suspense as to Mr. Glaisher's fate, when no response came to my questions," he wrote later. Finally Glaisher "gasped with a sigh."

"Do try, now do!" Glaisher remembered hearing as his pilot roused him from the deep, unrefreshing sleep. The instruments became dimly visible, and then Coxwell, and suddenly he was awake.

Coxwell remembered his confused look, but it passed quickly.

"I have been insensible," Glaisher said.

"You have," replied Coxwell, "and I too, very nearly."

Glaisher poured brandy on Coxwell's black hands, took up a pencil, and resumed his observations, reporting that "no inconvenience followed my insensibility." The scientist calculated later that during the time he was unconscious the balloon had reached about 37,000 feet, a height of roughly 7 miles, higher than anyone had gone before them.

The balloonists landed in a grassy field, and Glaisher walked more than seven miles to the nearest railway station to get assistance.

The *Times* of London exulted over the story. "We have just had an ascent such as the world has never heard of or dreamed of," its correspondent wrote on September 11. "Two men have been nearer by some miles to the moon and stars than all the race of man before them." The flight ranked among "the critical and striking moments of war, politics, or discovery," the newspaper stated. "But the feat was almost too audacious, and was carried on to the very verge of fate."

Glaisher would make 28 free-flight ascents in all between 1862 and 1866, although never again would he go so high. His observations proved valuable, most notably in testing rules proposed by Gay-Lussac in 1804—that temperature fell at the rate of 1 degree per 300 feet of ascent, and that the composition of the atmosphere is constant at all elevations. More generally, however, the findings of his daring expeditions were bound to disappoint the high expectations of scientific ballooning.

For the Frenchman Camille Flammarion, a veteran of several ascents, the first sight of manned balloon flight had evoked a "magnificent dream" about how the mysteries of the air would be laid bare. "Its secrets would be disclosed, the movements of the atmospheric world would be counted, measured, and determined as scrupulously as astronomers can determine those of celestial bodies; and man, once placed in possession of this terrestrial mechanism, would be able to predict rains and storms, drought and heat, luxuriant crops and famines, as surely as he can predict eclipses, and thus ensure an ever-smiling and fertile soil!" Coming to meteorology from astronomy, like Glaisher, Flammarion blamed scientists for the failure of the dream. "Is it not singular," he asked, that the science of the stars could foretell eclipses two centuries in advance, "whilst we can scarcely assert with probability what kind of weather we shall have tomorrow? The history of science tells us, however, that meteorological investigations have never been carried on with that energy and care which has long characterized the science of astronomy." Years earlier, Glaisher himself had written that he expected meteorology would "in due time rival the first of all sciences, astronomy."

Like many a new scientific instrument, the balloon illuminated a world of detail that posed more questions than answers. Glaisher's upper-air researches revealed an atmosphere that was more complicated and unpredictable than most researchers were prepared to imagine at the time. Rather than confirming old maxims, the new tool unexpectedly exposed a new layer of complexity, requiring new explanations. Gay-Lussac's estimation of the lapse rate of temperatures was confirmed only in the most general way, for example, because balloonists frequently encountered distinct layers of air of very different temperatures, humidities, and wind directions. As little as 500 feet off the ground, the wind could be blowing in the opposite direction from the surface. And air near the surface often was overlaid with warmer air at varying heights aloft.

Scientific balloonists also encountered more practical difficulties with their instruments than they expected. In the 1850s, Welsh recognized a problem with temperature observations caused by the fact that a rising balloon carries up with it a certain amount of air from lower levels of atmosphere. He rigged up an "aspirator," a bellows that forced ambient air into the thermometer, which seemed to work fine until especially cold temperatures were encountered. Glaisher took meticulous care of 22 devices that were part of the instrument kit he arranged on the table before him, but often he encountered difficulties with his data. For one, later observers decided that Glaisher had not taken sufficient precautions to avoid the effect of the sun's direct radiation on his thermometers. There were always surprises in the winds, a critical atmospheric feature that free-flying balloons were inherently unable to measure because they moved as part of these currents. And always there were anomalies, unexpected and unexplainable readings. Some were caused by the fact that the balloons moved up and down through the atmosphere more quickly than the instruments could effectively respond. A temperature reading taken at 10,000 feet on the way up could be much warmer than the temperature registering on the thermometer at 10,000 feet on the way down. Because free-flying balloons almost always shot up swiftly through the lower reaches of the atmosphere, balloonists were unable to get readings in the layers where much of the weather is formed. Glaisher and others sought to overcome this problem with a series of flights in captive balloons that were tethered to the ground.

The flights attracted increasingly widespread public attention, and the celebrity of the brave ballooning English scientist continued to grow. At the end of his aerial exploits, in 1869, Glaisher gave a series of lectures around England. None of this popularity had pleased Glaisher's boss at the Royal Observatory in Greenwich, the properly Victorian George Airy, Astronomer Royal. Airy disliked sensations of any kind, and meteorology was not a matter of great importance to him. Airy was among many astronomers of the age who distrusted meteorology's lack of theory. Glaisher had always felt more strongly about the value of meteorological work and was not afraid to pursue his own research agenda. In character, both men were strong-willed, arrogant, and autocratic. Within a few years of the balloon flights, during a disagreement with Airy over a small matter, Glaisher resigned from the observatory after 34 years as superintendent of the Magnetical and Meteorological Department. As he had with Airy, Glaisher had parted ways with Coxwell over some disagreement in the summer of 1864.

Recently Julian L. Hunt, a descendent of Glaisher, described his great grandfather as "a man of uncompromising and fixed opinions which were not always based on a calm assessment of the state of affairs." In the waning days of the reign of Louis-Napoléon, Glaisher was a staunch imperialist who resented the republican leanings of his French colleagues. Himself a man of humble origins and little formal education, he always looked down upon those of similar background in his professional and personal affairs. "He was obsessive both at work and in his personal relationships," Hunt wrote in 1996 in the *Quarterly Journal of the Royal Astronomical Society.* "One is forced to admit that Glaisher seems to have been a snob of the first order as far as his social and family life was concerned." In 1868, Glaisher's daughter, Cecilia, married a medical student, Frederick Hunt, from a family that Glaisher did not approve. "Mrs. Glaisher accepted the situation and her husband developed a coolness towards his wife and daughter which was never to be overcome," wrote Hunt.

For all that, Glaisher was a deeply committed scientist who made pioneering contributions to meteorology, including the organization in the 1850s of a network of volunteer weather observers in the British Isles that was a forerunner to the storm-warning system inaugurated by Robert FitzRoy in the 1860s. Studying ways to calculate the formation of dew, he would lay out on the cold grass during the night, an activity that brought on rheumatism, which caused him trouble for years. In the era before bacteria were known, he investigated the meteorology of London to determine if weather conditions played a role in the 1854 cholera outbreak. He helped develop mathematical "factor tables" which were used by scientists for years.

It was always the remarkable balloon flights for which he was most widely known, even though, for all their drama and daring, Glaisher did not achieve any real scientific breakthrough during those ascents. The British Association's interest in manned ballooning came to an end in 1881, when Walter Powell, a member of Parliament, was lost at sea in a balloon. But ballooning of that era opened the way to development in the 1890s of unmanned ballooning, which has a long and especially rewarding history in meteorology. At the turn of the century, the French meteorologist Léon-Philippe Teisserenc de Bort used unmanned balloons to discover the stratosphere, a region of relatively uniform temperature that Glaisher must have nearly touched in 1862. This new layer of atmosphere lent weight to an emergent idea about Earth as a system of concentric layers of fluids of different densities, a conception that eventually would become a fundamental principle in modern geophysical science. It would lead another meteorologist, the German Alfred Wegener, to discover the geological principle of continental drift. It was unmanned balloon flight that first disclosed depletion of the ozone layer. And unmanned balloons remain a mainstay of the daily upper-air measurements that help form modern weather forecasts.

PART II

AMERICAN STORMS

Good God! What horror and destruction!
—Alexander Hamilton, St. Croix Island, September 6, 1772

•

"It seemed as if a total dissolution of nature was taking place," young Alexander Hamilton wrote in a letter to his father of the hurricane that had raked the West Indies. Homeless families roamed the ruined streets, wounded and sick, "exposed to the keenness of water and air, without a bed to lie upon, or a dry covering to their bodies." As his account bears witness, nothing has stricken more lives with its terrifying power than Earth's life-giving and life-taking atmosphere. "The roaring of the sea and wind, fiery meteors flying about in the air, the prodigious glare of almost perpetual lightning, the crash of falling houses, the ear-piercing shrieks of the distressed were sufficient to strike astonishment into Angels."

The study of meteorology came naturally to North America, where land and sea contrive to provoke tempests great and small. Nowhere is weather more violent or various. While this fact of physical geography was not well understood in the nineteenth century, leading American scientists of the day would recognize that at least in the study of weather they held certain advantages over the more learned and better-established savants of Europe.

To British and European scientists, the appalling facts of hurricanes and tornadoes were unreliably observed details in improbable tales told by seafarers and fanciful literary descriptions in ancient texts. More generally, the European landscape just wasn't big enough to take in the shape and size of weather. Individual observers confined on little islands and narrow peninsulas were hard-pressed to grasp the hemispheric scale of events. Weather science needed a continent to work in.

The early decades of the nineteenth century saw a creative burst of groundbreaking work by Americans who were investigating the character of storms. Like the weather itself, the research led to rough-and-tumble debates that were typical of America at the time, a country young and beset with economic and social troubles. Central to these debates was the *American storm controversy,* but even after it died off like a spent cyclone, nasty political conflicts about the conduct and standards of American science would endure. It would take a civil war to settle the matter.

4

William C. Redfield

Walking the Path of Destruction

•

ONE OF THE FEW hurricanes ever to pass over New York City arrived at low tide on the evening of September 3, 1821. In one hour, the water rose 13 feet, swamping the wharves. Witnesses saw the tide from the East River actually meet the tide from the Hudson River across lower Manhattan. The eye of the storm ripped up the coast and came ashore over Long Island and passed over western Connecticut, Massachusetts, and Rhode Island. The rains and the winds were furious, felling trees across the countryside. New York City had never seen such damage from a storm. Elsewhere along the seaboard, the hurricane laid waste to the low-lying agricultural lands surrounding Chesapeake Bay. It drenched Philadelphia and caused widespread wind damage in the city. Hardest hit were the Outer Banks of North Carolina and the busy sea harbor of Norfolk, Virginia.

It came to be known as the Great Norfolk and Long Island Hurricane of 1821, but that name came later. At the beginning of the nineteenth century, storms went by a variety of names that were used interchangably. The distinguishing characteristics of such storms were among the many abiding mysteries of weather. Everyone knew there were differences, of course, but beyond their seasonal patterns, nobody could say exactly what they were. A storm was a storm was a storm. That they had recognizable structures, particular shapes, one kind from another, or that their winds behaved in certain ways, was not evident at the time.

In the early nineteenth century, a new generation of Americans would take up with enthusiasm and purpose the thread of ideas laid down by Benjamin Franklin many years earlier: that storms had structure and preferred directions of travel. It seemed to be an American problem, as it was a singular American experience to hunker down against winds of such incredible violence, praying to God that the family and the household survived this almighty threat of death and ruin. It would lead to a productive, golden era of weather research, in its own way. To the exclusion of almost everything else, the young science of meteorology in America was about to focus on just one question: *What is a storm?*

The Great September Gale of 1821, as it was first called, was the subject of more attention in scientific circles over the next few decades than any tempest that had hit the young country. It inspired debates that raged for years on both sides of the Atlantic. What made it so interesting was not the fact that it was so powerful, nor even the fact that it happened to hit New York, but the observations of its aftermath by a single, singular individual, and what they would come to mean to the science of weather.

A poor apprentice of a saddle and harness maker, a country merchant, an inventor, a naval engineer, and a New York transportation businessman, William C. Redfield was a man of uncommon intellect and industry. He was not a weather scientist, although he would make major contributions to understanding weather. He was not trained in a science of any kind, although he would contribute to paleontology as the first American expert on fossils of fish from the Triassic period, the time of the dinosaurs. And in 1848, he would become the founding president of the American Association for the Advancement of Science.

During a journey through the New England countryside shortly after the storm of 1821, Redfield happened to observe an interesting detail about the ravaged forests and fields across the landscape. It was there for anyone to see, of course, but it was not something that anybody had ever noticed before about the wreckage of such storms. In the apparent chaos was a large-scale pattern in the orientation of the trees blown down on September 3. Near his home in Middletown, Connecticut, the fallen trees were generally facing one direction, toward the northwest, and in western Massachusetts, less than 70 miles distant, the trees were facing the opposite way, toward the southeast.

Only someone of an especially astute and open frame of mind was likely ever to see it: in the pattern of debris was an imprint of the shape of the storm. In his mind's eye, a mystery was unfolding, as if some ancient nighttime marauder had finally left behind a set of distinct footprints. What Redfield envisioned was an enormous progressive whirlwind, a violent flow of air rotating counterclockwise around the axis of a center that was traveling from the southwest to the northeast with the prevailing winds.

However important he might have found these ideas, his old friend Denison Olmsted, a professor at Yale College, would later observe that Redfield was in no position to make them generally known at the time. In fact,

10 years would go by. "Amid all his cares," said Olmsted, "it clung to him, and was cherished with the enthusiasm usual to the student of nature, who is conscious of having become the honored medium of a new revelation of her mysteries. Nothing, however, could have been further from his mind than the thought that the full development of that idea would one day place him among the distinguished philosophers of his time."

William Redfield was born in 1789 near Middletown, the son of Peleg, a sailor who died at sea when William was 13 years old, and Elizabeth Pratt Redfield. A biographer describes his mother as "a woman of superior mental endowments." The eldest of six children, William at the age of 14 was apprenticed to a saddle and harness maker. His mother remarried, and in 1806 she moved to Ohio. The boy remained in Middletown, learning the trade. (Somewhere along the way he adopted the initial "C" to distinguish himself from two other William Redfields in the area, and he would later joke that it stood for "Convenience.")

When he completed his apprenticeship in 1810, Redfield set out on foot to visit his mother in Ohio, a journey of more than 700 miles. The trip took 27 days. So dense were the forests, so few the paths, that Redfield and two traveling companions skirted the shores of Lake Erie by walking along the beach. They covered an average of 32 miles a day. Passing through Albany, New York, the young men saw Robert Fulton's first steamboat, the side-wheeler *North River,* on the Hudson River. Redfield kept a journal describing the events of those days and the landscape they traveled. The long journey inspired a lifelong interest in transportation, and Redfield would use the experience years later to recommend the route of a railway between the Hudson and the Mississippi.

He returned to Middletown in 1811 and for 10 years took up the business of saddle and harness making and operated a store. In 1814, he married Abigail Wilcox. The couple had three sons, but the third newborn died in 1819, and soon afterward Abigail died from the effects of childbirth. In 1820, Redfield married his first wife's cousin, Lucy Wilcox. But a year later, childbirth claimed another son and another wife. Lucy died on September 14, 1821, soon after the great gale. It was in connection with this personal tragedy that Redfield and his eldest son, a small child, were traveling through the countryside of western Connecticut and Massachusetts, between Middletown and Lucy's family home.

"Few men have given more signal proofs of an original inherent love of knowledge," Olmsted said of his friend. The extent of Redfield's formal education was elementary and rudimentary.

He learned the sciences from the books in the personal library of a physician who settled in Middletown and from the influence of the Friendly Association, a debating society he helped organize. Throughout most of his life, Olmsted noted, Redfield's scientific knowledge was something he wrested from his days while other men were resting—by the flickering light of a fireplace as a boy apprentice, through the "dark scenes of domestic affliction and mournful bereavements" as the father of a troubled young family, and after a day filled with the concerns of business as an executive of a successful New York transportation company.

Redfield first became interested in marine navigation in 1820, when a fellow townsman, Franklin Kelsey, developed a new steam engine to propel a small boat. While the venture was unsuccessful, the experience launched Redfield into a career as a marine engineer. By 1822, he was operating a steamboat on the Connecticut River. In 1824, he moved to New York City and expanded his steamboat transportation business to the Hudson River.

Early steam engines were notoriously dangerous, and Redfield, as superintendent of a steam navigation company, set about to find ways to make them safer. He devised a system of passenger-carrying "safety barges" that were towed up and down the Hudson between New York and Albany by separate steam-powered tugboats. When the boats became safer, and the public became less fearful of steam power, Redfield converted the barges and started a large and profitable freight business. In New York, he was married a third time in 1828, to Jane Wallace, who survived him along with two sons by his first wife.

Olmsted met Redfield by chance—on the deck of a steamboat en route from New York to New Haven, Connecticut—10 years after the Great September Gale of 1821. Redfield introduced himself and asked Olmsted questions about a recent journal article he had written about hailstorms. And then he mentioned his own theory about the structure and behavior of the Great September Gale. Olmsted was impressed, not only with Redfield's ideas about the storm but with his careful presentation of the facts. He encouraged Redfield to write up his findings, which were published in 1831 in the *American Journal of Science and Arts*.

Redfield carefully described the direction of the winds, the timing of their arrival in different parts of the region, and the storm's apparent movement through New York and New England. He wrote:

> In reviewing these facts, we are led to inquire how, or in what manner it could happen, that the mass of atmosphere should be found passing over Middletown for some hours with such exceeding swiftness towards a point apparently within thirty miles distance, and yet never reach it; but a portion of the same, or a similar mass of air, be found returning from that point with equal velocity? And how were all of the most violent portions of these atmospheric movements, which occurred at the same portion of time, confined within a circuit whose diameter does not appear to have greatly exceeded one hundred miles? To the

> writer, there appears but one satisfactory explanation of these phenomena. This storm was exhibited in the form of a great whirlwind.

The rotating motion of some big tropical ocean storms had been recorded long before Redfield. In 1801, Redfield would later discover, Colonel James Capper of the East India Company described tropical storms off the coast of India as "whirlwinds whose diameter cannot be more than 120 miles." Capper wrote that "it would not perhaps be a matter of great difficulty to ascertain the situation of a ship in a whirlwind, by observing the strength and changes of the wind. If the changes are sudden, and the wind violent, in all probability the ship must be near the centre of the vortex of the whirlwind; whereas, if the wind blows a great length of time from the same point, and the changes are gradual, it may reasonably be supposed that the ship is near the extremity of it."

And an eighteenth-century seaman had left a famous account that similarly described a great rotary storm. "I was never in such a violent storm in all my life," wrote the English buccaneer William Dampier in 1687, detailing how his ship survived the passage of a "tuffoon" off the coast of China. "Tuffoons are a particular kind of violent storms," he said. He described the ghastly dark clouds, the sea afire with lightning, and the telltale pattern of winds. He noted the "falling flat calm" in the eye of the storm, and observed that "then the wind comes about to the S.W. and it blows and rains as fierce from thence, as it did before at N.E. and as long." Dampier was a pretty good naturalist, as pirates go, but he was no meteorologist, and no William Redfield.

In Redfield's 1831 essay, a violent and turbulent phenomenon of the atmosphere had been rendered coherent for the first time, shown to possess a particular dimension and structure and pattern of behavior. "Remarks on the Prevailing Storms of the Atlantic Coast of the North American States" was the first of a long series of contributions by Redfield to the *American Journal of Science and Arts*. In an article two years later, he described hurricanes and other storms of the North Atlantic in greater detail, identifying their origins in the Tropics and defining their parabolic paths first northwestward toward the U.S. coastline and then northeastward out across the ocean. Invariably, the winds circulate counterclockwise in the Northern Hemisphere, he noted, and accurately predicted that they would be found to circulate clockwise south of the equator.

Somehow, Redfield's paper quickly came into the hands of Lieutenant Colonel William Reid of the Royal Engineers, who began military duty in Barbados in 1831 soon after a devastating hurricane that year wrecked the island and killed 1,477 people in seven hours. Reid spent several years collecting numerous ships' logs and other reports of vessels in storms and presented his findings, along with Redfield's research, to a meeting in

London of the British Association for the Advancement of Science in 1838. The eminent English scientist Sir John Herschel congratulated Reid and said that he, too, had read Redfield's papers with great interest. He wondered if Reid had considered whether the warm Gulf Stream might account for the curious parabolic track of the storms' northerly progress over the Atlantic. Herschel lamented the fact that both researchers had avoided advancing a theory about the behavior of hurricanes. A theory, he said, "if it served no better purpose, helped memory, suggested views, and was even useful by affording matter for controversy, which might produce brilliant results, by the very collision of intellect."

Reid and Redfield forged a lifelong friendship and a fruitful collaboration. With access to Admiralty records, Reid supplied Redfield with data that permitted the pair to extend hurricane and tropical storm researches over the Atlantic and beyond.

This collaboration eventually led to the publication of handbooks that were used by seamen to avoid the worst hazards of ocean storms. In Britain, their research became part of the work published in 1848 by Captain Henry Piddington, *The Sailor's Hornbook of Storms in All Parts of the World*. In the United States, it became part of *The American Coast Pilot*.

In his report on his expedition opening Japan to the Western world in 1854, Commodore Matthew Perry said navigators were indebted to Redfield and Reid "for the discovery of a law which has already contributed, and will continue to contribute, greatly to the safety of vessels traveling the ocean." While other writers improved on the information provided seamen, wrote Perry, "to Redfield and Reid should be ascribed the credit of the original discovery of this undeniable law of nature and its application to useful purposes." A Redfield essay, "Cyclones of the Pacific," was published as part of the official documents of the Perry expedition to Japan. Redfield found that tropical storms and hurricanes of the western Pacific behaved just as those in the Atlantic.

The Redfield-Reid collaboration was remarkable for its depth and its selfless character, a quality that Perry had witnessed. In an introductory note to the essay on cyclones, the commodore praised Redfield for being "as remarkable for modesty and unassuming pretensions, as for laborious observation and inquiry after knowledge." Perry knew both Redfield and Reid, "and there is nothing more beautiful, as illustrative of the character of these two men, than the fact, well known to myself, that notwithstanding their simultaneous observations and discoveries, in different parts of the world, neither claimed the slightest merit over the other." Redfield and Reid maintained a lifelong correspondence and a deep mutual regard. The two men never met.

5

James P. Espy

"The Storm Breeder"

•

DESPITE THE EVIDENT care of William Redfield's investigations, his findings were sharply attacked, beginning in 1834, by a Philadelphia scientist with a very different theory about the character of storms. In an era of high seas sailing ships, when captains were relying on luck and lore for the safety of their vessels and men, the different ideas these two men held about the behavior of the winds were of special practical interest.

The air does not rotate around storms, James Pollard Espy insisted, but rather it rushes inward from all directions toward the center of low barometric pressure. Where Redfield described a whirlwind, a gravity-driven "aerial vortex," Espy envisioned a heat engine, a "chimney" of warm air expanding as it rises skyward, forming clouds and rain and sucking in surface air like some atmospheric black hole. Between Redfield's centrifugal theory and Espy's centripetal winds was a world of difference, or so it seemed at the time. Their methods of research were certainly different. Redfield's picture of the winds was the result of collating data from on-the-ground observation after a real storm. Espy's was laboratory work that led to a theoretical insight, the first description of the process of convection, and he spent the rest of his life trying to make every observation of every storm conform to it.

Like Redfield, James Espy came relatively late in life to the study of meteorology. He was born on May 9, 1785, in Westmoreland County, Pennsylvania, the son of French Huguenots. The family moved to the Bluegrass region of Kentucky while James was still an infant, but his father's Huguenot sensibility was so offended by the institution of slavery that eventually the family moved to the Miami Valley region of Ohio. Meanwhile, a daughter had married a Kentuckian from Mount Sterling, and young James stayed behind to be raised by his sister. When he was 18, he entered Transylvania University in Lexington, where he studied classical languages, graduating in 1808. Espy then followed his family to the frontier Ohio country, settling in Xenia, where he taught school and studied law. In 1812, he moved back east to Cumberland, Maryland, where he became principal of a local academy. There he married Margaret Pollard, thereafter adopting her last name as his middle name.

Apparently Espy did not take up any study of the atmosphere until sometime after 1817, when he moved to Philadelphia and became a private instructor of classical languages and mathematics at the Franklin Institute of the State of Pennsylvania. His first paper on the subject was prepared for the American Philosophical Society of Philadelphia in 1821, when he was 36 years old. It would be another 8 years before he would publish his first treatise on the subject of the influence of latent heat on the expansion of air—the core of his breakthrough on convection.

Espy and Redfield were as different from one another as their theories. Redfield was a successful businessman with a strong avocational interest in science. Espy had abandoned the field of classical studies and ever since was searching for ways to finance his meteorological researches. Where Redfield was diffident and shy, Espy was outgoing and egotistical and given to exaggeration. Where Redfield avoided public speaking, Espy was an eager proselytizer of his views.

Divided into opposing factions, the New York scientists supporting Redfield and the Philadelphia scientists in Espy's camp argued for years over the basic questions of the shape and character of storms. Across the Atlantic, Redfield was supported by eminent British scientists, while Espy won the day in France. A barrage of critical reports in scientific journals, polemic review articles, and emotional debates in meetings of scientists ran on for more than a decade. Professor Joseph Henry at the College of New Jersey in Princeton observed that "two hypotheses as to the direction and progress of the wind in these storms have been advocated with an exhibition of feeling unusual in the discussion of a problem of purely scientific character." It was, he said, "as if the violent commotions of the atmosphere induced a sympathetic effect in the minds of those who have attempted to study them." The battle was joined by a third protagonist, Robert Hare, a professor of chemistry at the University of Pennsylvania, who insisted that storms were phenomena of vast currents of opposing electrical charge. At a time when knowledge of electricity was fresh, many scientists were willing to speculate about what role it might play in the atmosphere, but only Hare seems to have taken his theory very seriously.

Redfield, the careful observer, began his descriptions by making no claims to having a theory about the cause of the storms or their motive force. The subject at hand was not yet *what causes storms* but more basically, and less ambitiously, *what are storms?* But Espy's elaborate theoretical approach to the subject, and his grandiose claims for it, seemed to force the issue. "A majority of readers are content with nothing short of a theory, or some hypothetical pretensions to one," Redfield wrote. When he finally took the plunge in this direction, he was not very persuasive.

Perhaps because of his engineering background, Redfield found a strictly mechanical explanation, what he called somewhat vaguely "the dynamics of the atmosphere." The atmosphere behaves like other fluids, he said, like water flowing over the uneven bottom of a stream or eddies of current whirling about an ocean. Its storms were whirls created by the force of gravity and by the effects of the rotating and orbiting Earth. Air pressure, temperature, moisture—everything else was secondary to the winds, he said. Air pressure was lowest in the center, he argued, because of the centrifugal force of the whirling fluids. A concave formation makes for less atmosphere over the center of a storm in the same way that water whirling in a basin forms a central depression. Redfield dismissed Hare's electricity as a by-product of storms and took direct aim at Espy, who had found his rotary winds "anomalous and inconsistent with received theories."

"The grand error into which the whole school of meteorologists appear to have fallen consists in ascribing to heat and rarefaction the origin and support of the great atmospheric currents which are found to prevail over a great portion of the globe," Redfield wrote. In his biographical memoir of Redfield presented in 1857 to the American Association for the Advancement of Science, Denison Olmsted lamented the fact that his friend had even made an attempt to advance a theory of storms, which he found vague and unsatisfactory. "I have almost regretted that he did not adhere to the ground he originally took, namely that he had not undertaken to explain the reason *why* the winds blow, but only to show *how* they blow," he said. "So far was matter of fact; all beyond was hypothesis. His facts are impregnable; his hypothesis doubtful."

James P. Espy's theory of storms was much more interesting and might have represented a striking advance in understanding the processes behind the formation of clouds and precipitation. When it came to his relations with fellow scientists, however, Espy was his own worst enemy. Even his friend Joseph Henry lamented a certain "want of prudence" in his colleague. Years later, another friend, Alexander Dallas Bache, would put his finger on a fundamental problem in Espy's approach to science. "His views were positive and his conclusions absolute, and so was the expression of them," wrote Bache. "He was not prone to examine and reexamine premises and conclusions, but considered what had once been passed upon his judgment as finally settled. Hence his views did not make that impression upon cooler temperaments among men of science to which they were entitled—obtaining more credit among scholars and men of general reading in our country than among scientific men, and making but little progress abroad." Espy had the habit of belittling the work of fellow investigators and so often offended other scientists. The controversy over storm theories took on a negative and bitter quality. Olmsted observed that large questions remained at the end of each theory of the causes of storms, and that "we still remain to a great degree in ignorance. Each of the combatants appears to me to be more successful in showing the insufficiency of the other's views than in establishing his own."

The core of Espy's conception relied on the law of gases that describes the relationship of pressure and volume, and on the release of latent heat by condensation. Espy applied English chemist John Dalton's interesting new idea that air containing more water vapor is lighter, rather than heavier, than air containing less. He employed the recent work of John Frederic Daniell on humidity and dew point and the insight of Joseph Black that water absorbs and releases latent heat in the processes of evaporation and condensation. For the first time, Espy accurately described the process of convection—the rise and expansion of warm air, its cooling to condensation, and its release of latent heat. Typically, he saw this insight in the grandest of terms: "It occurred to me at once that this was the lever by which the meteorologist was to move the world!"

Unfortunately, Espy saw as apparently essential to this vision of the energy behind storms the notion that wind rushes radially inward or straight toward the center of low pressure from all directions. When both Redfield and Espy examined the aftermath of the same storms, Redfield couldn't see any evidence for the winds turning inward, and Espy couldn't see evidence of any revolution. Neither Redfield nor Espy were aware of the work that was under way by the French mathematician Gaspard-Gustave de Coriolis—at the very time of the American storm controversy—which explained and quantified the effect of deflection on moving objects by the rotation of a sphere. As U.S. Weather Bureau meteorologist Eric R. Miller would observe a century later, Espy "could not admit the rotation observed by Redfield because he knew of no force capable of producing it." The seminal paper on what would become known as the *Coriolis effect* was published in 1835—too late to save Espy's pioneering theory of storms. In fact, the idea would not be taken up seriously by meteorologists for 25 years. Even if he had seen that article, he probably would not have changed his thinking. It was a hallmark of his character that once James P. Espy reached a conclusion, however improbable, he never looked back to critically reexamine his thinking. As one historian of science observed, so devoted was Espy to his findings "that he became a prisoner of his own theories." Everywhere he went, at every storm scene, in every lecture hall, he denied the rotary motion of tornadoes, waterspouts, hurricanes, and other tropical storms. Observations that suggested a whirlwind were just not done as well as observations made by James Espy. Like a prizefighter leading with his chin, he attached his important insight into cloud formation and precipitation to his mistaken and somewhat irrelevant idea about the flow of the winds.

Espy seemed to take quite literally the notion that convection was "the lever by which the meteorologist was to move the world!" He saw in convection the answer for everything about weather. "In short," he wrote, "it is believed that all the phenomena of rains, hails, snows and water spouts, change of winds and depressions of the barometer follow as easy and natural corollaries from the theory here advanced, that there is an expansion of the air containing transparent vapour when that vapour is condensed into water."

Espy claimed not only the ability to explain events, but to control them. Nothing made him more famous or more seriously damaged his reputation among fellow scientists than his grand assertion that he could move the world by making rain. He wasn't the first person to make such a claim, of course, but to the chagrin of his fellow scientists, he was the first meteorologist to do it. Before long, people who had no idea about the storm controversy among scientists had heard about Espy's claim that he could breed storms by setting fires. He became known as the Storm King, or the Storm Breeder.

"Few of our philosophers are better known to the public at large . . . and few names, even among political men, have found of late a greater circulation," observed a writer in the November 1841 edition of the *United States Magazine and Democratic Review.* "Who has not heard of his wonderful power of commanding the elements, of transforming the clearest summer afternoon into a cloudy, raining, thundering night, by ordering the invisible waters of the heaven to accumulate in a chosen place, and produce there clouds and rain, with all their usual consequences?"

Espy proposed an experiment: setting the western forests ablaze during the summer with weekly fires of 40 acres every 20 miles along a north-south line up to 700 miles long. He claimed that as a result of this conflagration "a rain of great length, north and south, will commence on or near the line of fires; that the rain will travel towards the east side-foremost; that it will not break up until it reaches far into the Atlantic Ocean; that it will rain over the whole country east of the place of beginning; that it will rain only a few hours at any one place . . . enough and not too much." Like everything else about his science, no amount of public ridicule would dissuade Espy from this view.

Frustrated with his inability to topple Redfield's work on winds, and running out of funds to finance his own research, Espy in 1837 began taking his science on the road with a series of public lecture tours. While this campaign of self-promotion brought financial rewards, it caused Espy to lose face among fellow scientists. More than one critic sneered that Espy, having failed to persuade fellow scientists of his views, had taken his propositions to a less discerning audience. And he certainly had. Public audiences enjoyed his earnest and artful presentations, and people who may not have understood his science certainly understood his rainmaking claims. He cultivated newspaper editors and reporters who fawned over his appearances. Writing in the *New York Express,* one enthusiastic journalist reported that "if professor Espy can do what he thinks can be done, *make a storm,* at once, man is almost master of the world."

In 1840, Espy took his cause to the savants of Europe, accepting invitations to address the British Association for the Advancement of Science and the French Académie des Sciences. Redfield also was invited, but he was unable to make the journey. Among the British, Espy encountered criticism of his theories from Sir John Herschel and Sir David Brewster. Their countryman, Lieutenant Colonel William Reid, Redfield's collaborator, already had presented the results of Redfield's researches to the organization, and Herschel read another Redfield paper after Espy's departure. In Paris, Espy found a more sympathetic audience. The mathematician François Arago was reported to have said, "England has its Newton, France its Cuvier, and America its Espy." Not everyone took this remark as a compliment to American science. A commission of the Académie des Sciences praised his work and urged that Espy be supported financially by the U.S. government.

Fresh from this remarkable accomplishment, Espy presented himself again to the U.S. Congress. He sought federal funds for a national system of simultaneous weather observations and funds for his own salary to oversee such a network. He also sought a provisional appropriation of $10,000 to be paid him, proportionally, if he were able to bring rain to 10,000 square miles during a drought. A visit from "Mr. Espy, the storm breeder," was recorded by John Quincy Adams, who, after serving as president, had become a member of the House of Representatives. "The man is methodically monomaniac, and the dimensions of his organ of self-esteem have been swollen to the size of a goiter by a report of a committee of the National Institute of France endorsing all of his crack-brained discoveries in meteorology," wrote Adams. "I told him with all possible civility that it would be of no use to memorialize the House of Representatives in behalf of his three wishes."

Arago may have been impressed with Espy, but the leading American mathematician of the day, Benjamin Peirce of New York, was not. Peirce wrote in a critical article in 1842 that he had attempted to be impartial but probably had failed, "for there is an air of self-satisfaction and contempt for the views of other observers in his statements, which irresistibly arouses the demon of obstinacy. Even storm-kings are intolerable in a republic."

A more sympathetic Joseph Henry, who as secretary of the Smithsonian Institution often found himself trying to mediate the controversy, observed after Espy's presentation to the British scientists at Glasgow that it was a "matter of congratulation" that the savants of Europe were taking such notice of all of the meteorological research under way in the United States. "The interesting theories of Espy and of Redfield, contradictory as they may now appear, will probably be found not incompatible with each other," he wrote, "and they undoubtedly form the most important steps towards the widest generation which have yet been attempted in reference to the complex phenomena of the motions of the atmosphere."

6

Elias Loomis

Mapping the Storm

•

ON FEBRUARY 4, 1842, a tornado raked through Mayfield, Ohio, and across 24 miles of northeastern Ohio farmlands and woods before disappearing over Lake Erie. Fast on the scene was Elias Loomis, a professor of mathematics and natural philosophy at Western Reserve College in nearby Hudson. Loomis was no ordinary investigator. At the age of 30, his reputation for meticulous research was already well established among the combatants of the American storm controversy. No shaky theory, no unsupported hypothesis, no mere conjecture was safe in his presence. As a student at Yale College, Loomis had studied under Professor Denison Olmsted, William C. Redfield's friend and scientific benefactor. Quiet and self-absorbed, Loomis was a young man of independent mind and method, and his work on the storms of February 1842 would make him famous. Along the way, there was the question of the deplumed fowls to clear up.

"In the tornado which occurred at Stow in 1837, a circumstance was remarked which I had never seen noticed before, that several fowls were picked almost clean of their feathers," he reported in April 1842 to the Connecticut Academy of Arts and Sciences. "In the New Haven tornado of 1839, the same fact was noticed. I made particular inquiry on this point at Mayfield."

What Loomis was looking for were clues to the velocity of the winds of the tornado. It was a tricky question. Some objects seemed to have been lifted with miraculously benign force. So it was that an 11-year-old boy attempting to close a door against the wind was lifted out of a window and deposited uninjured a distance that Loomis measured by chain as 14 rods from his house. When the boy looked back, the house was gone. In other instances, lightweight objects such as pieces of clapboard siding were imbedded in the ground as deeply as two feet—as if shot from a cannon.

"What charge of powder is capable of producing the same effect?" Loomis asked. He had some earlier calculations by another scientist to rely on, but, of course, "I desired to verify it by experiments of my own." He charged a cannon with one and a quarter pounds of powder behind a six-pound ball, pushed in some short pieces of oak board, and fired at a nearby hillside. Then he measured exactly how deeply the wood shards were imbedded in the ground. "We arrive then at the conclusion that the clap-boards at Mayfield were driven into the earth with a velocity of 1,000 feet per second, or 682 miles per hour."

So what velocity of wind is required to pick a fowl clean of its feathers? James Espy had repeated to Loomis the story of a man who had seen a tornado strip chickens and turkeys completely of their feathers, leaving the birds otherwise uninjured, although they soon died. After asking around Mayfield, Loomis reported many dead birds in the wreckage, their wings and legs broken, and a "flock of hens was carried off in the tornado, and they have not since been seen." Only one goose and a turkey was found to be deplumed by the wind. Still, Loomis wondered what velocity of wind would do such a thing. Besides, the use of a gun for meteorological research is not always conveniently at hand. Science was not to be denied.

Loomis charged the cannon with five ounces of gunpowder, and instead of a ball, he stuffed in a freshly killed chicken.

> As the gun was small, it was necessary to press down the chicken with considerable force, by which means it was probably somewhat bruised," he reported. "The gun was pointed vertically upwards and fired; the feathers rose twenty or thirty feet, and were scattered by the wind. On examination they were found to be pulled out clean, the skin seldom adhering to them. The body was torn into small fragments, only a part of which could be found. The velocity is computed at five hundred feet per second, or three hundred and forty one miles per hour. A fowl, then, forced through the air with this velocity, is torn entirely to pieces; with a less velocity, it is probable most of the feathers might be pulled out without mutilating the body. If I could have the use of a suitable gun I would determine this velocity by experiment. It is presumed to be not far from a hundred miles per hour.

Loomis attempted another innovation while researching the Mayfield tornado. Espy and Redfield had invariably found confirming evidence of their own contradictory theories by studying the same evidence—the compass bearings of fallen trees. Even more than a hurricane, the sudden and extreme violence of a tornado makes an especially random and chaotic track of the debris. Loomis thought he had contrived the crucial experiment to ascertain the sequence of the tornado's winds by measuring the orientation only of trees that were stacked on one another. The bottom tree fell first, he reasoned, and the top tree fell last. While no one was entirely convinced of this new approach, Loomis concluded that there was evidence in the fallen trees of both radial and rotary winds. "That some tornadoes are whirlwinds certainly cannot be denied," he wrote. "That the motion at right angles to a radius is sometimes quite small compared with the centripetal motion, seems equally clear." But he found "infinitely improbable" the idea that the rotary motion in a tornado "should ever become mathematically nothing."

Loomis was not satisfied with these results. The Mayfield tornado was part of a much larger winter storm that swept the continent, one which Loomis was determined to study in great detail. His research would do more than shed new light on the rival theories of the American storm controversy. Loomis would define a much larger atmospheric phenomena that did not conform to either Redfield's or Espy's models.

If the American storm controversy was going to be resolved, clearly it wasn't going to be resolved by the principal antagonists. Redfield and Espy seemed preoccupied with attacking each other in print. The potential for brilliance that Sir John Herschel had seen in controversy over rival theories had not been realized. What the dispute lacked was what the study of weather phenomena had always lacked. What was needed was not more theories, but enough reliable data to accurately test them. Without instruments of measurement distributed widely across the countryside, weather phenomena remained the subject of conjecture. So public and so bruising had become the controversy that the dispute had settled into a battle of wills. Their scientific activities had come relatively late in the lives of both Redfield and Espy, and so they had little personal experience with the give-and-take of it, and there was little by way of scientific convention in the young nation to guide them.

Espy's activities in particular—his public lecturing and travels, his petitioning of Congress and his self-promotion—looked a lot like a political campaign, and in some sense it was. At a time when the young nation was facing a widespread economic crisis and science was a new idea, he was seeking political support for federal funding for meteorology, and not incidentally for a job for James Espy. Meanwhile, whenever a tornado or other storm would pass through, partisans from both camps would rush out with tape measures and compasses and find just the kind of evidence they expected. What midcentury meteorology needed was more observations, more reliable observations, and a fresh face.

The scientist who would do most to clear the air would be Elias Loomis, who had watched the storm debate unfold in the 1830s during his years of postgraduate tutorship at Yale. Loomis was one of the most accomplished scholars of his day, and his investigation of American storms would influence meteorology long after the Redfield-Espy controversy had been forgotten. Ironically, Loomis would be most remembered not for what he found out about the storms he studied, but for the graphical method he used in presenting the results of his work.

Loomis would follow Redfield in his conservative scientific method, meticulously gathering and collating facts until a pattern eventually emerged. This was the approach to science that he would follow throughout his life, and his research would be famous not for its brilliance but for its accuracy and the reliability of its detail. He was not a man of intuitive theoretical insight. But if there was a theory or hypothesis to test, Loomis was the man to do it.

With a little bit of luck, and a little more theoretical bent, he might have helped resolve the American storm controversy sooner. After three years of tutoring at Yale, Loomis accepted a position as professor of mathematics and natural philosophy at Western Reserve College in Hudson, Ohio. Preparing himself for that role, he was allowed to spend the first academic year, in 1836 and 1837, studying in Europe. Loomis purchased equipment for his new laboratory and spent most of his time in Paris, attending the lectures of such great mathematicians and physicists as François Arago, Jean-Baptiste Biot, and Siméon-Denis Poisson. As it happened, it was during 1837 that Poisson applied to the motion of artillery shells Gaspard-Gustave de Coriolis's calculations of the deflective effect of Earth's rotation. In time and place, Loomis was in close reach of the mathematics that might have brought a merciful conclusion at least to the part of the ornery debate having to do with the behavior of winds in hurricanes and other large storms. If Loomis had heard those lectures by Poisson and brought that theory with him back to the United States, he might have been able to show both Redfield and Espy why the winds flowing toward low pressure were bound to rotate. In any event, the American controversy would grind on.

Instead, Loomis had something else important to show Redfield, Reid, Espy, and others who sought evidence that would confirm their theories from the observations of more and more storms. He adopted the opposite approach. "I resolved to select some single storm of strongly marked characteristics, and trace its progress as extensively and minutely as possible," he wrote in an article read to the American Philosophical Society in Philadelphia in the spring of 1840. He was expecting to test the wind field in a way that would clarify the ongoing Redfield-Espy dispute, but instead he found himself examining a type of storm that did not fit either theory very well.

For his purposes, he chose the big winter storm that raked across the United States during the week before Christmas in 1836. He chose this storm because he had available an unusually large number of observations. Sir John Herschel, who at the time was at the Royal Observatory at the Cape of Good Hope in South Africa, recently had proposed that meteorologists and volunteers everywhere undertake hourly observations of barometric pressure, temperature, and other conditions for a period of 36 hours around the winter and summer solstice and the vernal and autumnal equinoxes that year. Herschel's plan won wide acceptance. Fortuitously, a big winter storm was raking the North American continent on December 20, 1836, and Herschel's scheme provided Loomis with the best data anyone had ever compiled for any storm. Loomis had hourly readings from 8 cooperating eastern stations from Quebec to Baltimore, barometer readings from 27 stations, and other information from much of the country east of the Rocky Mountains, as well as from Bermuda, the West Indies, and from a ship along the Pacific Coast.

Nobody had ever undertaken the kind of painstaking study, or even realized the global character, of such a storm. It ran the length and breadth of the country and beyond, north into Canada and south into Mexico. He mapped the storm at six-hour intervals, recording air pressure, temperature, wind direction, velocity, and precipitation. On a map of the United States, he drew lines showing the advance of the wave across the country.

In size and structure, this storm was very different from the tropical storms and tornadoes that had been the subject of most investigation and theorizing. Loomis found himself defining what meteorologists long after him would call a *midlatitude cyclone,* or extratropical cyclone: a storm with a long atmospheric wave structure of hemispheric reach that looks and acts very differently from the tornadoes and hurricanes and other tropical storms that Redfield, Reid, and Espy had been fighting over. There was no theory for Loomis to use to frame the observations. It would be seven decades before a German scientist would confirm almost exactly the phenomenon of a cold current underrunning warm air. Nowhere could Loomis find Redfield's rotary winds, mainly because the low-pressure center of the storm was north of all observing stations. Loomis found opposing winds rushing toward one another, as Espy would have supposed, but he came up with a very different picture of what was causing this phenomenon.

He raised this question: "How is it possible for two winds, not far separated from each other, to blow violently towards each other for hours, and even days, in succession?" And he answered it: "The conclusion is inevitable; the north-west wind displaces the south-east one by flowing under it." Loomis drew a cross-sectional diagram of what modern meteorologists would immediately recognize as a cold front storm. The cold air mass is digging along into the warmer air, which is pushed up off the surface, forming clouds and precipitation as it rises. Still, Loomis was not satisfied with these results.

In his 1890 memoir of Loomis, Professor Hubert Anson Newton, a Yale College astronomer and mathematician, observed that with the big map showing the progress of the low-pressure wave Loomis had made some advance in the graphical presentation of meteorological information. "But even the method he then used was entirely unfitted to give answers to the questions which meteorologists were asking," he wrote.

Those questions were posed in circulars issued by the joint committee on meteorology of the Franklin Institute and the American Philosophical Society. Espy chaired the committee, and the questions, as framed by Newton, reveal the state of American weather science in the middle of the nineteenth century: "What are the phases of the great storms of rain and snow that traverse our continent? What [are] their shape and size? In what direction, and with what velocity do their centers move along the surface of the earth? Are they round, or oblong, or irregular in shape? Do they move in different directions in different seasons of the year?"

The breakthrough came in February 1842. Loomis had not only the Mayfield tornado to investigate, but two intense winter storms that swept across the United States from the 2nd to the 5th and from the 15th to the 17th. Unlike the 1836 storm, the centers of the winter storms of 1842 were within the area of the observation network, allowing Loomis to draw a more complete picture of their structure, especially the behavior of the winds.

His report on these storms, read by Alexander Dallas Bache at the centennial meeting of the American Philosophical Society on May 26, 1843, is one of the most famous documents in meteorology. "On Two

Storms which Occurred in February, 1842" not only shed important new light on the American storm controversy, but created an immediate sensation and a lasting change because of Loomis's innovative graphical presentation of his data.

The report marked the first publication of what came to be known as a *synoptic weather map*—lines on a map connecting *like values* for measurements of air pressure, temperature, wind, sky conditions, and precipitation. The term applies to a weather map covering a large area, although for meteorologists *synoptic* has also come to mean *simultaneous* measurements over a large area. The *isochronal lines* that connect like values observed at the same time looked remarkably like the kinds of maps geologists drew to represent geomagnetic values, Loomis's other great scientific preoccupation, but nothing like these 13 charts had ever before been published in meteorology.

As Loomis himself observed, "Nearly every circumstance essential to a correct understanding of the phenomena of the storm is thus presented to the eye at a single glance." The depiction of data from 131 observers on 13 charts was more than a matter of convenience or style. For the first time, meteorologists really could see the features of the atmosphere that constitute the dimensions of a storm and their relation to one another. Such a synoptic weather map would become basic to weather prediction. Newton called it the "greatest of the services which our colleague rendered to science."

This new method of graphical presentation would forever change the way meteorologists depict the features of storms. It would render for the mind's eye the information that makes possible a dynamic picture of a moving storm. The spatial representation of those contour lines are the basis of the simplest modern weather maps and the visual output of some of the most sophisticated computer models.

Loomis's findings were based on much larger and more complex storms than either the Redfield hurricane model or Espy's theory, which best described a thunderstorm. There was no doubt in his mind that much of the rain and snow fell as a result of the process of convection described by Espy. But Loomis's vision of the cause of the storm was far ahead of both Espy and Redfield, and far ahead of his time. Already, in the 1836 storm, Loomis had anticipated concepts introduced seven decades later that saw these storms as conflicts between air masses along frontal boundaries. In 1842, he was using terms that became central in even more modern views of winter storms as atmospheric waves moving over the globe.

Not only did Loomis define the structure and behavior of a midlatitude winter storm for the first time, but he tackled two fundamental questions that no other meteorologist of his day would approach: "First, what caused the formation of the first cloud . . . ? And secondly, why should the storm once organized ever cease?" The cause he ascribed to "the superior weight and density of the air in the central parts of the United States," which was the aftermath of a storm two days earlier, and his mid-February storm left behind a similar dense cold air mass that provoked the clouds of yet another storm. "Thus, one storm begets its successor," he said. "The undulations thus excited in the atmosphere bear considerable analogy to the waves of the ocean agitated by a tempest, and which are propagated by mechanical laws long after the first exciting cause has ceased to act." For the cause of the storm's demise, he pointed to the charts showing the wave of cold air advancing on the center of the storm. "This cold north-west wave advanced more rapidly than the center of the storm," he wrote. "It was beginning to counteract the increase of temperature arising from the condensed vapour, and after that the violence of the storm must rapidly decline."

At the time, of course, the big question among meteorologists had to do with the winds. Loomis showed the diagrams of Espy's radial winds and Redfield's circling winds and pronounced neither of them fit to explain what happens with the big storms of winter. "The accompanying charts will show that neither of these diagrams faithfully represents the storms here investigated," he said, "and it is doubtful whether either of them ever accurately represents the motion of the wind over any large portion of the earth's surface. . . . A combination of the two is, however, frequently seen." Loomis not only showed the inwardly spiraling wind field but explained its cause. While the winds might "be expected to flow inward towards the center of the storm," he wrote, in a large storm "such a central tendency, whether in air or water, uniformly causes the fluid to approach the center, not exactly in radii, but circuitously." He showed the increasing rotation of the winds on the charts from one day to the next. "There is a physical cause for this rotation, and for its being uniform in direction, in the case of large storms," he wrote. "The southerly wind has a greater motion eastward than the parallels upon which it successively arrives, arising from the rotation of the earth; and the northerly wind a less motion eastward, for a relative motion westward." This meant that the winds were always going to blow the same way around low pressure in the Northern Hemisphere. As Loomis put it, "in this region the circulation, in great storms, is probably invariable in direction."

In this and later papers—the ones not published by Espy—Redfield's outline of circulating winds more closely followed these new contour lines on the map linking areas of like air pressure, which would come to be known as *isobars,* around the center of a storm. Espy never overcame this problem. "The earnest and deep convictions of the truth of his theory in all its parts, and his glowing enthusiasm in regard to it; perhaps, also, the age which he had reached, prevented Mr. Espy from passing beyond a certain point in the development of his theory," Bache would later observe. "The same constitution of mind rendered his inductions from observation often unsafe." Because Espy could never bring himself to adapt his theory to the new wind data, his more fundamental contribution to the understanding of cloud formation and precipitation through the process of convection fell into obscurity.

Concluding his famous paper, Loomis suggested that if more investigations were carried out along the lines he had shown:

> [W]e should soon have some settled principles in meteorology. If we could be furnished with two meteorological charts of the United States, daily, for one year, charts showing the state of the barometer, thermometer, winds, sky, etc., for every part of the country, it would settle for ever the laws of storms. No false theory could stand against such an array of testimony. Such a set of maps would be worth more than all which has been hitherto done in meteorology. Moreover, the subject would be well nigh exhausted. But one year's observation would be needed; the storms of one year are probably but a repetition of those of the preceding.

7

Joseph Henry

Setting the Stage

•

ELIAS LOOMIS'S GROUNDBREAKING display of contemporaneous measurements from several points across the country made plain to the small scientific community toward the middle of the nineteenth century the crucial importance of a widespread network of observers. Cooperative, coordinated, far-flung observations and a central organization are uniquely indispensable to the science of meteorology. Without a large field of vision the science of storms is nearly blind. As James P. Espy put it: "The astronomer is, in some measure, independent of his fellow astronomer; he can wait in his observatory till the star he wishes to observe comes to his meridian; but the meteorologist has his observation bounded by a very limited horizon, and can do little without the aid of numerous observers furnishing him contemporaneous observations over a wide-extended area." Without observations from an area at least as large as the event they intended to study, investigators had good reason to wonder whether the science would ever answer the questions it had posed about the nature of storms.

Loomis called for a "general meteorological crusade" to replace the "guerrilla warfare which has been maintained for centuries with indifferent success." In his report on the 1842 storms, he argued that a "well arranged system of observations spread over the country would accomplish more in one year, than observations at a few insulated posts, however accurate and complete, continued to the end of time."

The "few insulated posts" of the U.S. Army had been taking weather observations as early as 1819. Army surgeon general Dr. Joseph Lovell had begun this practice in the hope of finding a link between weather conditions and the health and effectiveness of soldiers. These reports were of little meteorological interest, however, until 1842, when Espy was hired by the Army Medical Department and began trying to use the system for the study of storms.

While the participants and close followers of the storm theory debate pined for a large observation system, to the country at large, in the 1830s and the early 1840s, there was little to recommend such a scheme. The nation was in deep economic trouble. In Hudson, Ohio, Loomis's pay was in arrears at Western Reserve College, and money was so scarce that much of business across the country was being done through bartering. In Andrew Jackson's America, politicians bragged about their commonness and humble roots. The social value of science in general was vague, at best, and the value of meteorology was especially unproved. Beyond the theoretical disputations of the few ardent combatants in the American storm controversy, about all that meteorology had to offer society were James Espy's grandiose claims that he could breed rainstorms by setting large fires.

Loomis held out the hope that if enough citizens would enlist in his weather observation crusade, "the war might soon be ended, and men would cease to ridicule the idea of our being able to predict an approaching storm." But he was really offering more than he could deliver in 1843. Without the means to communicate quickly over great distances, the discovery that the great storms of winter blew fairly reliably across the country from west to east was merely interesting. There was no reason to suppose that a large weather observation network could be used to foretell the movement of storms from one place to another across the land in time to actually issue warnings in advance of their arrival anywhere.

The proposition that future weather could be foretold was not something that was discussed seriously at the time in polite company. In England especially, the very idea was scandalous. Predicting future weather still had not outlived its long and unsavory association with divination, medieval astrology, and witchcraft. There were laws against such things. On the Continent, where the savants jealously guarded the sciences within their conservative societies, France's preeminent astronomer and physicist, François Arago, could confidently declare in 1845: "Never, no matter what may be the progress of science, will honest scientists who care for their reputation venture to predict weather."

Only extraordinary circumstances in America—an unlikely coincidence of events and the support of a special individual—would carry the idea of a continent-wide weather observation system beyond the status of a scientific daydream. The money fell like manna from heaven, in the form of a curious bequest from an Englishman. And fortunately for the science of meteorology, it fell into the hands of Joseph Henry. This great physicist, the most famous American scientist of his day, would do more than secure financial support for meteorology at a pivotal moment. From the laboratory research of this same individual would come the technical means to transform the science.

Along with Sir Charles Wheatstone and Michael Faraday in England, Henry in the 1820s and 1830s was discovering the principles of electromagnetism that underlay the use of electrical power. At the College of New Jersey in Princeton, Henry was entertaining students with an electromagnetic telegraph 10 years before Samuel F. Morse patented such a device and brought it to market. "In the discovery of the mode of magnetizing soft iron at a distance by means of currents of galvanism, and in his invention of this little machine, was not merely the possibility, but the fact of the electro-magnetic telegraph," wrote a former student, Professor Henry D. Cameron of the College of New Jersey. Henry was a research scientist, unselfishly devoted to the principles of science, and that meant to him that he was not in the business of making practical applications of his work. In 1845, soon after the first telegraph line was strung between Washington and Baltimore, Henry was quick to see what it could mean to meteorology. A method of timely long-distance communication would bring new purpose to a network of observers across the land and dramatically elevate the importance of weather science in general.

Nobody knows what possessed James Smithson, the illegitimate son of the duke of Northumberland, to bequeath his fortune to a nation he had never set foot in, providing for a "Smithsonian Institution" to be built in Washington, D.C., for the "increase and diffusion of knowledge." Nobody had ever heard of such a thing. Smithson died in 1829, although the bequest did not take effect until 1836 when his only heir, a nephew, died childless. In a time of national recession, Congress haggled for 10 years over what to do with the $508,318.46 before an independent board of regents took office in 1846. As they searched for a director, letters recommending the professor of mathematics and natural philosophy at the College of New Jersey in Princeton came from leading scientists, including Arago in France and Faraday and Sir David Brewster in England. "The mantle of Franklin has fallen upon the shoulders of Henry," wrote Brewster, drawing a comparison that would often be applied to Henry. Observers would recognize in both men a common scientific acumen, liberal philosophical bearing, and moral stature. A man who had never attended college, Henry was the Smithsonian regents' unanimous choice as first secretary of the institution.

Weather scientists knew they had a good friend in the new secretary of the Smithsonian. Like Franklin, Henry had devoted hours of his days over years of his life to the study of the nature of electricity, including lightning, and the mysteries of the atmosphere. He made no claims as a meteorologist, although he had a powerful grasp of the subject and wrote extensively about meteorology over the years. He was a good friend of Espy's and one of the most interested and informed observers of the controversies of weather science swirling around him.

Early in his career, Henry had become personally familiar with the work of one of the first small-scale networks of volunteer observers to be organized in the United States. In 1825, the state legislature had charged the regents of the State University of New York with putting together such a system statewide. Two years later, Henry, a young new professor of mathematics and physics at Albany Academy, was hired to help organize the data. For the five years he remained at the academy, Henry studiously tabulated these observations and issued annual reports, supplying information that was useful more for climate than for weather studies.

Espy did similar work for the Commonwealth of Pennsylvania under the auspices of the meteorological committee of the Franklin Institute in Philadelphia. In 1837, the Pennsylvania legislature appropriated $4,000 for instruments for a system of volunteer observers. Conflicts arose with officials of individual counties who were made responsible for the instruments and observers, and the fragmented network was discontinued after about 10 years.

Loomis tried but failed to persuade the state legislature to finance a similar system in Ohio.

While planning the Smithsonian's first major undertaking, a national network of meteorological observers, Henry took Espy's and Loomis's advice and relied on the vision Loomis developed in his 1842 appeal to the American Philosophical Society to underwrite such a program. "The United States are favourably situated for such an enterprise," Loomis wrote. "Observations spread over a smaller territory would be inadequate, as they would not show the extent of any large storm. If we take a survey of the entire globe, we shall search in vain for more than one equal area, which could be occupied by the same number of trusty observers. In Europe there is opportunity for a like organization, but with this encumbrance, that it must needs embrace several nations of different languages and governments. The United States, then, afford decidedly the most hopeful field for such an enterprise." His appeal to national pride must have been especially attractive to Henry and the regents, who were eager to make a reputation for a new national American scientific institution. As Loomis put the case: "There are but few questions of science which can be prosecuted in this country to the same advantage as in Europe. Here is one where the advantage is in our favour. Would it not be wise to devote our main strength to the reduction of this fortress?"

Loomis was as quick as Henry to realize the potential value of the infant telegraph system to meteorology. "When the magnetic telegraph is extended from New York to New Orleans and St. Louis," he wrote in a letter to Henry, "it may be made subservient to the protection of our commerce."

Reporting to the Smithsonian regents in 1847, Henry said that American science had recently made more contributions to meteorology than any other physical science. He added: "Several important generalizations have been arrived at, and definite theories proved, which now enable us to direct our attention with scientific precision to such points of observations as cannot fail to reward us with new and interesting results. It is proposed to organize a system of observations which shall extend as far as possible over the North American continent." The time for such a network was "peculiarly auspicious," he observed, because "citizens of the United States are now scattered over every part of the southern and western portion of North America, and the extended lines of telegraph will furnish a ready means of warning the more northern and eastern observers to be on the watch for the first appearance of an advancing storm."

The Smithsonian launched two weather science programs. The first was the creation of a continent-wide network of observers who daily recorded meteorological observations and mailed them every month to the secretary in Washington, where they were tabulated, published annually, and made available to researchers. Within a few years, this system included approximately 600 regular observers and effectively spread the name of the institution and fueled public interest in weather science. To be a "Smithsonian observer" was a mark of some civic distinction. The data collected by these observers was used extensively by meteorological researchers for many years.

The second, smaller program proved to be the more publicly popular and quickly changed political thinking about the value of meteorology to the United States. In 1849, Henry launched a network of about 20 telegraph operators scattered across the nation. These station operators would routinely activate their operations each day with a sign-on signal that briefly described the weather in their locales. By 1850, a large map of the United States depicting the weather conditions across the country was displayed in the lobby of the Smithsonian Institution. Simple symbols of different colors showed where it was raining or snowing and where it was clear. They also graphically illustrated the eastward progress of large winter storms. The big weather map became a favorite exhibit among visitors to the Smithsonian, who were always interested in seeing what weather their distant friends and relatives were experiencing. Henry could occasionally be seen at the map explaining to visitors the principles of weather and the movement of storms. Public interest continued to grow. Henry forwarded a copy of the Smithsonian weather map to the *Washington Evening Star* for daily publication. He became so confident in the patterns of storm movements he discerned that when rain or snow appeared in Cincinnati, he would publish in the *Evening Star* a cancelation of Smithsonian lectures that night.

The Civil War was about to wreak havoc on the Smithsonian's network of weather observers, as it did on many national institutions. A disastrous fire struck the Smithsonian in 1865 and the cost of restoring the ornate structure would cripple its ability to rebuild the meteorological program. But the institution had achieved its goal in any case—advancing the science of meteorology and the "diffusion of knowledge" so demonstrably that the federal government would create a national weather service in 1871.

To Cleveland Abbe, who for 45 years would be the chief scientist of the new weather service, there was no doubt of Henry's role in its development. "Although our country can boast of many able meteorologists, who have greatly promoted our knowledge of the laws of atmospheric phenomena, it is safe to say that to no single worker in the field is our nation more indebted for the advancement of this branch of science in its present standing than to Joseph Henry," Abbe wrote. "To him is undoubtedly due the most important step in the modern system of observation—the installation of the telegraph in the service of meteorological signals and predictions."

Henry was secretary of the Smithsonian Institution until the end of his life. Suffering from nephritis, a kidney disease, at the age of 80, and surrounded by loved ones and anguished friends, Joseph Henry died on May 13, 1878. With his last words, he asked about the weather, about which way the wind was blowing. His funeral and memorial services were the kind of public spectacles the capital reserves for great national leaders.

8

Matthew Fontaine Maury

A Storm of Controversy

•

LIEUTENANT MATTHEW FONTAINE MAURY of the U.S. Navy conceived and executed a great but relatively simple idea: collect the logbooks of vessels crossing the ocean, plot their courses and wind observations on charts, then publish and distribute the charts to all the seafarers who sent in their logbooks. *Charts of the Winds and Currents* was the first systematic information about the best routes to take across the ocean and what conditions to expect, on average, at different times of year. For the first time, the ship captains of the world were sharing the benefits of their experience. These charts were credited with shortening voyages and saving money and lives at sea. They won Maury an international renown. This great navigational advance was considered by many people to be a work of great science, and so they took Maury to be a great scientist. First among them in this respect was Matthew Maury.

The man who did so much for nineteenth-century navigation was widely respected for his achievements and well-connected politically for a time, but he was never able to convince the leading American scientists of his day that he was one of them. They fought him for control of meteorology, among other sciences. To men like Alexander Dallas Bache and Joseph Henry, Maury was not a scientist either by training or by practice. He was in all respects a practical military man of large vision and boundless energy and ambition. He was a prolific writer and a popular public speaker on sundry scientific topics, although he was neither a scientific observer nor a theorist. Much of what he wrote was about the scientific work of others, and yet he was notorious for failing to acknowledge these debts. Matthew Maury's name, and only Matthew Maury's name, appeared on papers and articles.

Still, he occupies a curious cultural divide. Now as then, different and contrary estimations of Matthew Fontaine Maury inhabit the literature. Most historians and scholars of the humanities who have written about Maury uncritically describe him as a scientist, even as a man of scientific genius, as the founder of the science of oceanography. To historian Bernard de Voto, writing a century later, Maury was "a universal genius" and the nation's "foremost scientist" of his day. Maury's popular book *The Physical Geography of the Sea,* published in 1855, often has been referred to as the founding textbook on the subject of oceanography, especially by people who have not read it.

The Physical Geography of the Sea is not a scientific text but a series of unorthodox commentaries, lyrically written and bountifully laced with scriptural references, about marine and atmospheric phenomena. Many of its generalizations were discredited and outdated even at the time they were written. *The Physical Geography of the Sea* was never used as a textbook, unless it was used by Maury himself at the end of his career when he was a professor at Virginia Military Institute. Then, as now, most scientists of the sea and the weather did not regard Maury as the founder of their sciences. To the French physicist Marcel-Louis Brillouin, writing at the turn of the century, *The Physical Geography of the Sea* already was a curious artifact. "On every occasion his religious optimism breaks forth in lyrical paragraphs of a rather naive inspiration but of fine literary form," the Frenchman wrote. "In addition to his qualities as a writer, Maury had a practical, if not scientific mind." Like other scientists, Brillouin drew the critical distinction between Maury's meteorological speculations and his practical achievements with navigational charts. "This success, a direct result of the observations collected and having no relation whatever to Maury's views concerning the circulation of the atmosphere, must have created the illusion that the latter were valid." While Maury certainly drew popular attention to these fields, his works are not and were not subjects of serious study among scientists. His name does not appear in their literature.

Born in 1806 near Fredericksburg, Virginia, raised in Tennessee and educated in its rough country schools, Maury had joined the navy as a midshipman in 1825. He spent nine years at sea, first aboard the *Brandywine* when it returned the Marquis de Lafayette, the Revolutionary War hero, to France after his triumphal tour of the United States. Maury then traveled around the world on a four-year cruise in the *Vincennes,* and later he was part of an extended cruise off the Pacific coast of South

America. These voyages impressed him with the lack of navigational instruction for naval officers. In 1834, he wrote a book on marine navigation that eventually became a standard naval text. In 1839, a stagecoach accident permanently disabled his right leg and brought his sea duty to an end. This injury led to his appointment in 1842 as superintendent of the navy's Depot of Charts and Instruments in Washington, where he launched his big chart project. But the ambitious Maury always had more than hydrography in mind.

Although he was untrained and unpracticed in astronomy, Maury was nevertheless considered by the secretary of the navy, fellow Virginian John Y. Mason, to be the best man in the country to become superintendent of the new U.S. Naval Observatory in 1844. While Mason probably was bound to appoint a naval officer rather than a more qualified civilian scientist, there were better astronomers in the navy to choose from, and other officers with stronger scientific credentials. This appointment, along with Maury's grand ambitions for himself and the institution, brought him into growing conflict and competition with the leading American scientists of the 1850s, especially the inner circle led by Alexander Dallas Bache. Physicist Joseph Henry, Harvard's Benjamin Peirce, the leading mathematician of his day, and the great Swiss-born geologist Louis Agassiz were part of a group that called itself the Lazzaroni, a self-deprecating allusion to the beggars of Naples. They were the leaders of a generation of university-trained scientists that had emerged since the turn of the century, and they were out to elevate the professional standards of American scientists and the stature of American science. To these scientists, especially Henry and Bache in Washington, Maury was a charlatan in their midst, all the more a threat for the political power he could wield.

In Washington, where former president John Quincy Adams was one of the few members of the House of Representatives interested in science, politicians of the era were easy prey to entrepreneurial inventors and engineers and pseudoscientists who were always angling for federal sponsorship of their schemes. It was a time when expeditions were organized to test the theory that there were "holes in the poles" that gave access to Earth's interior. James Espy was promoting the idea of setting forest fires in the West to bring rains to the East. As early as 1838, Henry was writing to Bache, who was traveling in Europe, about "a most disgusting form of charlatanism" that led the Senate committee on naval affairs to hail certain "ridiculous and puerile" claims about terrestrial magnetism. Henry called the congressional report that described the claims as worthy of public confidence and congressional patronage a "disgrace to the country," and protested official publication of such claims before they were tested by scientists. "I often thought of the remark you were in the habit of making that we must put down quackery or quackery will put down science," Henry wrote.

To Henry and Bache, Maury was an even more direct threat. At the U.S. Coast and Geodetic Survey, Superintendent Bache lived in fear that the navy would succeed in its efforts to wrest control of his civilian agency from the Treasury Department and make it part of Maury's Hydrographic Office at the observatory. With Maury's backing, the navy's effort almost succeeded in 1851. Behind the scenes, Bache and Maury fought for control of various surveying programs along the Atlantic coast and on the high seas. In the nation's capital, they were competing power brokers, fighting over the principle of military or civilian control of American science as well as access to the purse strings of the federal government.

Along with his navigational charts, Maury issued ever-enlarging sets of *Sailing Directions and Explanations* in which he expounded his iconoclastic views on a wide range of scientific matters. Readers of these commentaries saw no mention of the work of others: the ocean and coastline work of the Coast Survey, for example, or the meteorological research sponsored by the Smithsonian. Here was Bache, the great-grandson of Benjamin Franklin, enduring the effrontery of Maury waxing disputatiously on the cause of the Gulf Stream and invoking the research of Bache's illustrious ancestor to support his views. Contemporary scientists saw little of value in Maury's explanations of the great current, except the need for more research, and later oceanographers saw even less. To Henry Stommel, a famous twentieth-century oceanographer and an expert on the Gulf Stream, Maury was "very much confused concerning fluid mechanics, even though that science was being rapidly advanced in his time."

By 1851, there appeared to be little in the fields of science that Matthew Maury did not seem ready to call his own. However many employees he was able to put on the payroll of the observatory, nobody else seemed to accomplish anything worth writing about under their own names. When the civilian astronomer Sears C. Walker quit after briefly working at the observatory and took his research to be published by the Smithsonian, Maury was furious and demanded the return of the observatory's property. (Walker went to work for Bache.) So it was Maury alone who was writing "On the Variable Light of Clio" at one moment and, at the next, Maury alone who was writing "On the Probable Relation between Magnetism and the Circulation of the Atmosphere." In

August 1851, when Bache stepped down as president of the American Association for the Advancement of Science in Albany, New York, there was no mystery who he was trying to warn his colleagues about when he alluded to the ambitions of certain pretenders: "Our real danger lies now from a modified charlatanism, which makes merit in one subject an excuse for asking authority in others, or in all, and because it has made real progress in one branch of science, claims to be an arbiter in others."

By 1851, Maury had ventured far beyond his expertise as a maritime navigator and continued to expand his horizons. While the Smithsonian Institution already was collecting and displaying meteorological observations from its own large and growing network of volunteers across the country, Maury was staking claim to the science of weather across the countryside as well as the sea. It was only a matter of time until Matthew Maury, with the help of the Old Testament, would explain everything. With nothing more substantial than the ambition to do it, the naval officer began telling farmers that he could accomplish for agriculture what he had achieved for commerce and navigation at sea.

When the British Royal Engineers asked the United States if it would be interested in developing a joint system of land-based meteorological observations, Matthew Maury seized on a grander idea. Although there was no agreement about meteorological observations in and around the United States, Maury began promoting the idea of international cooperation in worldwide uniform meteorological observations on land as well as the sea. His call for the convening of an international conference to secure a system of standard weather reporting schemes failed to win the support of leading scientists in the United States or in Europe. The resulting International Conference on Maritime Meteorology, held in Brussels in 1853, was reduced in scope to standardizing meteorological observations by naval vessels at sea. With the exception of the host, Belgian mathematician and astronomer Adolphe Quételet, the meeting was attended not by scientists but by naval officers.

Returning to the United States, Maury very soon plunged into writing *The Physical Geography of the Sea*. The book owed its existence not to Maury's urge to instruct seafarers, as with the navigation text, but to a warning from his publishers that his other writings accompanying his charts were not protected by copyright. It reads like it was written by a man in a hurry. With its florid oratorical style and its constant reliance on Scripture, it reads like the sermon of a southern preacher of the day. Still, its poetic bombast lured many uncritical readers. Appearing first in 1855, the book was published in several editions and remained in print for 20 years in the United States and in England, despite its scientific fantasies.

"Maury brought in hypothetical electric and magnetic forces when his dynamical courage failed," Sir Napier Shaw would observe in 1924. A modern critic, the Harvard scholar John Leighly, noted that the book "never received the approval of those contemporary scientists who were best able to judge it . . . and can not be looked upon as representative of the best scientific thought" of the 1850s. Moreover, Leighly wrote in 1963, "the history of the application of physical theory to the sea and the atmosphere would have been little different if the book had never been written."

Maury at this time also redoubled his campaign to take over land meteorology in the United States. Soon he was abandoning all pretense of cooperating with Henry. The power struggle came to a head on January 10, 1856, when Maury spoke to the U.S. Agricultural Society at the Smithsonian Institution. Maury was pushing for passage of a bill in Congress that would appropriate $20,000 to support a land meteorology system according to Maury's plan and under Maury's direction. In the audience was Joseph Henry, who had learned of the land meteorology plans in the newspaper.

Maury said that the time had come when "observations on the land are found to be essential to a successful prosecution of our investigation into the laws which govern the movement of the grand atmospherical machine. At sea we have the rule, on the land we look for the exceptions. We want to see the land, therefore, spotted with co-laborers observing also according to some uniform plan, and such as may be agreed upon in concert with the most distinguished meteorologists at home and abroad." But as far as Maury was concerned, the plan was going forward whether or not the most distinguished meteorologists at home and abroad liked it.

Henry rose to speak, proclaiming his appreciation for Maury's achievements with his navigational charts, according to the society's journal of the meeting:

> This, however, said Mr. H., though of great importance, was in itself a simple matter, and involved no great scientific attainment, and certainly implied no particular ability to apply a meteorological system for the seas to the land for agricultural purposes. The methods and objects are entirely different. The latter problem is one of much complexity, and, in order to devise a proper system, the subject should be submitted to a commission of the scientific men of the country who had paid particular attention to meteorology and to the problems connected with agriculture.

Declaiming any rivalry, Henry drew a sharply barbed distinction between the two men. Whereas everything he did was in the name of the Smithsonian Institution and according to the bequest of John Smithson, "on the contrary, whatever Lieut. Maury has done is in his own name, although at the expense of the government." Now the animosity was out in the open. By some accounts, the illegitimacy of James Smithson's birth came up. Maury and his allies prevailed at the meeting in 1856, but Henry and his allies prevailed in Congress and saw the bill defeated in 1857. Maury then took leave from his position and began a 25-city speaking tour, urging popular support for his land meteorology plan. But his influence on the subject would end with the outbreak of the Civil War in 1861, when he resigned his position at the observatory to join the Confederacy.

In the heat of battle in 1861, a report in the *Washington Star* recorded the depths of his fall from grace in the nation's capital. "The Humbug Maury turned up again," the paper reported, and was said to be inventing submarine batteries to protect Southern harbors. "In the city of Washington Capt. Maury is so thoroughly appreciated that no invention purely of his own will be likely to receive much consideration," it went on. "His career, when those qualified to do it justice shall undertake to depict it, will exhibit one of the most remarkable and successful careers of unblushing charlatanism known in the world's history."

Maury may have been in eclipse as a national figure, but the country was not through with men willing to make large claims about their knowledge of weather. Describing himself as a "certified practical meteorologist," one enterprising individual gained access to President Abraham Lincoln in the middle of the Civil War and claimed that his ability to forecast the weather would save "thousands of lives and millions of dollars." It was the kind of claim that Matthew Maury might have made had he been born a Northerner rather than a Southerner. But Frances Capen made the kind of mistake Maury never would have committed. Trying to impress the president with the value of meteorology to warfare, Capen issued a weather forecast, and as a result his career as a presidential adviser was over before it began. According to the noted weather historian, the late David Ludlum, Lincoln wrote on April 28, 1863: "It seems to me that Mr. Capen knows nothing about the weather, in advance. He told me three days ago it could not rain again till the 30th of April or 1st of May. It is raining now [and] has been for ten hours. I can not spare any more time to Mr. Capen."

9

William Ferrel

A Shy Genius

•

IF THERE WAS A TRUE intellectual giant among the meteorologists of the nineteenth century, a man who was both able and inclined to devote scientific genius to the problems of the atmosphere, a man who could define the behavior of the air in such a way as to set the science on a whole new plane, a Newton or a Kepler of meteorologists, it was a certain American country boy who learned his physics with a pitchfork while carving geometric diagrams in the soft poplar of a barn. That boy would be William Ferrel. Although great astronomers and physicists had taken up problems of the atmosphere from time to time, adding insight and advancement here and there before turning back to their principal endeavors, Ferrel's was the first really powerful intellect to focus sustained attention on meteorology.

Unfortunately for the science, this Irish American was born into such poverty and such raggedly low rural circumstances that his scientific education came especially slowly and unevenly, and recognition of the importance of his work, when it came at all, arrived especially late. Contemporaries described his "extreme diffidence" and complete indifference to personal position. While these self-effacing characteristics may have endeared him to fellow researchers, they very likely contributed to the obscurity and late blooming of his reputation. So it happened that one of the most important papers in the history of meteorology was written by an unknown Tennessee schoolteacher and was first published in 1856 in the *Nashville Journal of Medicine and Surgery.*

Ferrel's epoch-making explanation of the general circulation of the atmosphere appeared first as "An Essay on the Winds and the Currents of the Ocean" and then, with more mathematical development, as "The Motions of Fluids and Solids Relative to the Earth's Surface," which was published in 1859 and 1860 as successive chapters in *Runkle's Mathematical Weekly.* "This alone, had he written nothing else, would have assured his fame in after years, when it would have been discovered," wrote a friend, Professor Frank Waldo of Boston, the author of an early influential text on meteorology. Neither the Nashville journal nor the *Mathematical Weekly* were likely to circulate among meteorologists, and the influence of these papers was slow to materialize in the United States and in Europe, especially in Britain, during the second half of the nineteenth century. Among men who understood what they were reading, however, Ferrel's essays had an immediate and profound effect. The eminent American meteorologist Cleveland Abbe never forgot the experience. "They gave me at once the strong conviction that a successful attack had at last been made on the complex mechanics of the atmosphere, and that ultimately all would be unraveled," he wrote. "I have often said that that memoir is to meteorology what the 'Principia' was to astronomy." The allusion was less extravagant than it might seem, for Ferrel was a celestial mechanic in the tradition of Pierre-Simon Laplace and Sir Isaac Newton. There was what Abbe called "an intellectual inheritance."

William Ferrel was born "of humble parentage," as he called it, in 1817 in the backcountry of south-central Pennsylvania, the eldest of six boys and two girls. His father, Benjamin, was in the lumber business for a time and had an interest in a sawmill. The rural schools William attended "were of a very inferior order, the teachers mostly being able to teach only reading and writing and a part of the arithmetics," he recalled. "It was thought that I made very rapid progress and I soon had the reputation of being the best scholar in the neighborhood." When William was 12, his father bought a farm in Berkeley County, Virginia, now part of West Virginia. The move there in the spring of 1829 did nothing to improve his educational opportunities. "I was now kept closely at work on the farm, but I went to school here two winters and completed my common-school education," he wrote. "The school-house was an average one of the country at that time—a rude log cabin with oiled white paper instead of glass for window panes." And after school there was nothing but farmwork, "nothing to engage my mind in the way of study and nothing to read" except a weekly newspaper. Somewhere he ran across an advanced arithmetic text, and with the first 50 cents he earned, by harvesting at a nearby farm, he traveled to Martinsburg to purchase the book. In the flickering light of a fire or a tallow candle flame, soon he mastered the work, his intellect fired especially by the brief sketch and few diagrams on geometric measurement.

On the morning of July 29, 1832, while on his way to the field, Ferrel witnessed a solar eclipse. "This excited me and set me to thinking," he wrote. "I had not read or thought anything about astronomical subjects before, but knew somehow that an eclipse of the sun was caused by the moon's passing between it and the earth, and that the lunar eclipses were caused by the moon's passing through the earth's shadow." Working only with a calendar and an elementary geography text, 15-year-old Ferrel set to work to create his own solar ephemeris, accurately describing the times of solar and lunar eclipses. "The amount of study I gave to the subject, both day and night, was very great," he remembered. His work was finally done in 1834. During the next year, he accurately predicted one solar eclipse and two lunar eclipses; on the later occasions, he noted, "the greatest error in the predicted times was only nine minutes."

The farm and the country were poor in those days, and the Ferrel family was big. Most everything William really wanted had to be borrowed or bartered for. In February 1834, he went looking in Martinsburg for a treatise on trigonometry and instead came away with a surveying text, which served his purpose even better. "It happened that during the summer I was engaged a great part of my time on the threshing-floor, which had large doors at both ends, with wide and soft poplar planks," Ferrel wrote. "Upon these I made diagrams, describing circles with the prongs of the pitchfork and drawing lines with one of the prongs and a small piece of board." The diagrams on the doors remained visible for more than a quarter of a century, he said, "and in returning occasionally to the old homestead I always went to take a look at them." In the middle of the winter of 1835–1836, he rode for two days to obtain a copy of *Playfair's Geometry* in Hagerstown, Maryland. In 1839, with money earned from teaching, he finally enrolled in a small college in Mercersburg, Pennsylvania, where he got his first look at algebra. But his money ran out before he could complete his courses of study. It wasn't until July 1844, at the age of 27, with financial help from his father, that he graduated from little Bethany College in Virginia.

Ferrel happened upon a copy of Newton's *Principia* in a country store in Liberty, Missouri, where he was teaching school. Studying this work, he recalled, "I now became first interested in the tides and conceived the idea that the action of the moon and sun upon the tides must have a tendency to retard the earth's rotation on its axis." Then he obtained a translation of Laplace's five-volume *Mecanique Céleste* and delved more deeply into this subject. This is what Cleveland Abbe meant by an intellectual inheritance: from Newton through Laplace there was a straight line to William Ferrel's first scientific paper, published at the age of 35 in *Gould's Astronomical Journal* in 1853. "On the Effect of the Sun and Moon upon the Rotary Motion of the Earth" was the first to point out what mathematicians call a "second order" discrepancy in Laplace's formulation of the frictional effect of the tides on Earth's rotation.

In a bookstore in Nashville, Tennessee, where he was now teaching, Ferrel picked up a copy of Matthew F. Maury's *The Physical Geography of the Sea,* which he read and found to be scientifically flawed in many important respects. "This first turned my attention to meteorology," Ferrel recalled. As the eminent Harvard scientist William Morris Davis would later observe, "This almost gives reason to be obliged to Maury for putting his theory in so convincingly impossible a form." Still later writers would remark that turning Ferrel's gaze onto meteorology was Maury's greatest contribution to the subject. Ferrel told a friend, Dr. William K. Bowling, the publisher of the *Nashville Journal of Medicine and Surgery,* that "I did not agree with Maury in many things," although he declined Bowling's invitation to write a critical review of the book. Ferrel abhorred the kind of controversy that seemed to so animate men like Maury or James P. Espy. Instead, in response to this request, he wrote his first meteorological paper, "An Essay on the Winds and the Currents of the Ocean," a famous work that mentions Maury's views only incidentally.

In Ferrel's hands, for the first time, the deflective force of Earth's rotation finally was given its appropriately fundamental place in shaping the behavior of the global winds and the currents of the ocean. Not since the trade winds researches of Edmond Halley in 1686 and George Hadley in 1735 had the subject of the general circulation of the atmosphere been considered with such care. Ferrel gave the atmosphere both motion and shape—bulging in the middle latitudes, at about 35 degrees, where the force of Earth's rotation deflects to the east its poleward-leaning upper winds and to the west its equatorial trades at the surface. With this work, and the more rigorous treatment of the subject a few years later, Ferrel established himself, along with Laplace, as a founder of geophysical fluid dynamics, a rigorous interdiscipline that is the mother of modern earth sciences.

In the hands of this country boy, the American storm controversy finally came to an end. "In those days it required close study and a very clear insight into the subject and a very independent mind to prevent one from becoming a partisan either of Maury or of Espy or of Redfield, not to mention Dr. Hare, the electrician of Philadelphia, or Dove, the statistician of Berlin," Abbe wrote later. "Precisely such a mind we find in Ferrel, who, from his surroundings in Nashville, was enabled to look out upon the world of meteorological disputants and to insist upon logic and reason in place of a war of words."

William C. Redfield was nearly right about the rotary motion of winds around hurricanes and most other storms, of course, and Espy was wrong about them blowing in from all sides directly toward a center. Still, he dismissed Redfield's effort to explain the gyration as having to do with "the peculiarities of the aerial currents" and Maury's vacuous suggestion that it "may be owing to the magnetism of the air." Ferrel's model of the general circulation of the atmosphere not only explained the rotary shape of the big storms as a function of the effect of Earth's rotation, it made it absolutely essential. On the other hand, wrote Ferrel, Espy was right about the source of the energy of such storms. Hurricanes, "and all ordinary storms, must begin and gradually increase in violence by the action of some constantly acting force, and when this force subsides, friction brings the atmosphere to a state of rest," he wrote. "This force may be furnished by the condensation of vapor ascending in the upward current in the middle of the hurricane, in accordance with Professor Espy's theory of storms and rains."

In 1857 came an invitation to join the staff of the *American Ephemeris and Nautical Almanac,* an annual publication of the U.S. Naval Observatory, in Cambridge, Massachusetts. Ferrel left Nashville a schoolteacher and arrived in New England a man of science. After 25 years at home on the farm, after nearly 15 years of grammar school teaching, Ferrel began the professional career for which in every way he seemed to be intended. At the age of 39, William Ferrel had finally found his calling. More accurately, his calling had found him. He never asked for a job, and many of his important papers were written not from his own impulse but at the urging of friends. Until his retirement in 1886, at the age of 69, he devoted his life to science, to studying the mechanics of the tides and to meteorology and, in a loosely knit pattern, to elaborating his thinking about the planetary motions of the atmosphere and the oceans that he first characterized in 1856.

Ferrel's scientific articles, even the "popular" treatments which contained no equations, were different than others on the subject. As fellow researcher Alexander McAdie put it, Ferrel "built a science, and lifted meteorology from a mass of observations and description, with a few self-appearing general laws, to an exactitude requiring the most refined methods of mathematical discussion." He was a brilliant deductive theorist. Abbe wrote:

> Many of our colleague's papers are to be recognized as successful efforts to solve problems that had hitherto been considered beyond our reach.

> So completely did he examine his field of study, so accurately did he select the important from the unimportant forces at work in nature, that in every one of the numerous results formulated by him we recognize that a distinct advance has been made from which there will be no need to retreat in future years. In the study of such complex phenomena as the motions of the ocean and air we have to consider the presence of some eight or ten factors, either one of which may become at times of preponderating importance. At some distant epoch man may be able to effect a general solution of equations that shall express the simultaneous influence of all these diverse forces, but at the present time we are not able to even write out the equations, much less to resolve them.

If Ferrel ever took the time to look at a thermometer or a barometer or make any other atmospheric observation himself, the occasion was not recorded anywhere. "He does not seem to have been at all blind to the occurrence of external facts," wrote Davis, "but he appreciated that the proper understanding of meteorology must be based on wider observations than could be made by any one person." For his data he relied on others, such as the meticulous storm studies of Elias Loomis, the global wind observations compiled by James Coffin, and the facts about barometric pressure contained in Maury's *Physical Geography of the Sea*.

Ferrel was not a great mathematician, but he was, like Newton, good enough. Davis called Ferrel's ideas in meteorology "wonderfully original" and far more important than his mathematical treatment of them. "It is perhaps because of too great attention to mathematical form and relative neglect of the idea that it clothes that the English mathematicians and meteorologists as a whole have been so little affected by Ferrel's suggestions," Davis wrote in 1891. "His principles as yet have not really touched meteorological science in that conservative country."

The situation in British meteorology in those years was ironic at the least, if not scandalous. Old men of dimming vision were in charge of the scientific societies. To borrow a phrase from one historian, there were dead hands at the tiller. At a time when the English mathematicians and scientists were so sententiously berating the applied, empirical work of the pioneering forecaster Robert FitzRoy, they were themselves cultivating a fairly thorough ignorance of some of the nineteenth-century's most important theoretical work in meteorology. For years, British meteorologists were repeating old theories long after they had been abundantly disproved.

Ferrel himself seemed largely indifferent to these matters. "In Ferrel's everyday life there seemed to all who knew him to be a certain amount of isolation too large to satisfy us, and he himself felt this," Abbe wrote of his old friend. "Preoccupied by problems, and therefore necessarily alone with his own thoughts, he had, even from boyhood, neglected the development of the social side of his nature." He never married, never lost that painful shyness, passing his days living with friends in what Abbe described as "a systematic regularity from which he knew that he must not depart if he were to master the problems of the ocean and the air." A few minutes of parlor conversation with people he knew well was all the social interaction that he required, and all that he really seemed comfortable with. "Of the leading traits of my character a prominent one all through my life has been a great diffidence and a backwardness in coming in contact with strangers or in putting myself forward in any manner," Ferrel wrote. This trait was to cause him trouble even at the height of his scientific powers. He described how on at least one occasion he could not bring himself to stand before a sympathetic assemblage of the American Academy of Sciences in Boston and read a paper, "Note on the Influence of the Tides in Causing an Apparent Acceleration of the Moon's Mean Motion." He wrote, "Although the paper contained an original and important suggestion and I had it all written out, yet I carried it to the meetings of the Academy time after time with the intention of reading it, and my courage failed, and if I had deferred it one time more I would not have anticipated [Charles] Delaunay on the same subject."

In 1867 came an offer of a job at the Coast and Geodetic Survey, which was interested in his insights into tidal observations. At the survey, he developed a tide-predicting machine. Beginning in 1884, this elaborate device was used for years by the agency to compute tides mechanically.

In 1882, while still overseeing the construction of his tide machine at the Coast and Geodetic Survey, Ferrel accepted an invitation to work as a professor of meteorology at the new U.S. Army Signal Service Office, where his longtime friend Cleveland Abbe had been in charge of weather forecasting for the nation since 1871.

For 30 years Ferrel conducted his highly original and valuable research, producing 3,000 pages of studies that other investigators would pour over for many years. For all that, he ended his own brief description of his career by expressing a sense of lost opportunity in his life. "Much of my time has been wasted," he wrote, "especially the earlier part of it, because not having scientific books and scientific associations I often had nothing on hand in which I was specially interested."

William Ferrel died on September 18, 1891, in Maywood, Kansas, in the company of his brothers and sisters. In the following months, in all of the important societies in the country, men of science gathered to pay their respects.

"Here was a man known by name to hardly more than a few hundred of our millions; known personally to fewer still in a vast population that is ever ready to recognize notoriety; and yet his quiet work greatly advanced the bounds of human knowledge," wrote Davis. "It is a curious commentary on renown to name Ferrel, of whom the great world knows nothing, as the most eminent meteorologist and one of the most eminent scientific men that America has produced."

Cleveland Abbe, who knew Ferrel longer and better than probably anyone, never lost the feeling that he was in the presence of someone who was in some important way intellectually different from other men. "We all remember his quiet ways, his indefatigable industry, his shyness, his perpetual absorption in the contemplation of some new and complex problem," he wrote. "He lived in an atmosphere of abstraction; he was with us, yet not of us."

PART III

The Main Artery

•

While modern consumers of the science certainly view the daily forecast as meteorology's main product, if not its whole point of existence, hardly more than a century ago it was nowhere near certain that weather prediction was even an appropriate goal. In the middle of the nineteenth century, leading British scientists were scandalized by the thought; and for several decades, on both sides of the Atlantic, conflicts over the place of weather forecasting gave meteorology a terribly fractured look.

"The universal desire for information about future weather opens the main artery of communication between the science and the public and is the chief vindication of an appeal for public funds," observed Sir Napier Shaw in 1924. Even at that late date, however, Shaw was questioning whether British meteorologists, who were among the last to permanently establish official forecasting in 1879, might have moved too quickly: "They might have given their attention to more purely scientific aspects, with some assurance of useful results, because the scientific aspect is the aspect from which true knowledge is derived."

Throughout the era, meteorology was marked by conflict between two very different cultures and this apparently irreconcilable divide: "practical" weather forecasting on one side, "pure" scientific research on the other. As early as the 1860s, wherever scientists were in charge, as they were in Britain, they suppressed popular forecasting. As late as the 1920s, wherever forecasters were in charge, as they were in the United States, they quashed promising research.

10

Robert FitzRoy

Prophet Without Honor

•

THE FIRST NATIONAL weather forecaster was an unlikely man in an unlikely place. He was Robert FitzRoy, a vice admiral in the Royal Navy, who in 1831, as captain of the HMS *Beagle,* had selected a young scientist named Charles Darwin as the naturalist to accompany him on a surveying expedition to the coast of South America. This eminently practical man became in 1854 the chief statistician of the new Meteorological Department of the Board of Trade. Like Joseph Henry across the Atlantic, FitzRoy very quickly realized how the new electromagnetic telegraph could be used to warn Her Majesty's seamen and fishermen of the threats of advancing storms. He began issuing predictions of coming storms early in 1861. In pioneering this service, he was fulfilling a vision that was far ahead of many people in England. So far ahead, in fact, and so ridiculed and criticized for this forwardness by the reigning savants of the Royal Society, so ardently sure and frustrated was he, and finally so sick with despair, that at 7:45 A.M. on April 30, 1865, he walked into his bathroom and slit his throat with his razor. There his family found him dying.

In the aftermath, FitzRoy's detractors in Parliament and in the ivory towers of British science covered themselves in a lengthy, highly detailed, and essentially unfair report about the accuracy and value of his storm warnings and weather forecasts. They were bound and determined to return meteorology to the realm of theory, like physics and astronomy, to the philosophical pursuit of knowledge for its own sake. As far as the savants were concerned, collecting data for this cloistered research was the very purpose that FitzRoy's statistics gathering had been intended to serve. Instead, however, this navy man had gone off in a completely different and unauthorized direction. Weather forecasting was empirical—just too practical, too unorthodox, and really too imprecise to be worthy of the word *science*. That something might actually be owed the British public by way of such a service was not a very large part of the Victorian order of the day. A special committee of the Royal Society recommended to the Board of Trade that it discontinue issuing storm warnings and daily weather forecasts until further notice, and so, on December 7, 1866, the British government complied.

As the Royal Society and the government would discover, however, the tragedy of Robert FitzRoy would haunt British science for years. It would change the shape of meteorology as a science and undermine the scientific establishment's ambition to secure public approval for a long-term government endowment for "high science." At the end of his life, pushing with characteristic zeal for a system of weather forecasting for the benefit of the British public against the resistance of a government bureaucracy and a scientific elite, the first national weather forecaster had history on his side. Eventually, of course, the Royal Society would have to abandon its stand against weather forecasting and the renamed Meteorology Office would take up where it left off. The invention of the telegraph and the advent of the weather map had put in the hands of "practical men" the tools to anticipate the arrival of dangerous weather, however imperfectly, and there was no putting that genie back in the bottle.

Even before he took the job at the new Meteorology Department, Robert FitzRoy was a troubled man, a perfectionist with an unstable temperament and a history of controversy and personal and professional disappointment. Because of his temper, the captain had earned the nickname "Hot Coffee" from the crew of the *Beagle*. Having dined daily with the captain for five years, Darwin had more than once witnessed his splenetic outbursts. Driven by work and worry, fearing for his own sanity, FitzRoy had suffered at least one breakdown during the voyage. "I never cease wondering at his character," Darwin confided later to a friend, "so full of good and generous traits but spoiled by such an unlucky temper. . . . Some part of the organization of his brain wants mending." The two men were far apart in temperament, politics, religion, and science, although the great naturalist seemed always to retain a fondness for his sea captain.

An aristocrat in bearing and background and a staunch Tory, FitzRoy was elected a member of Parliament from Durham in 1841 and appeared to have a promising political career before him. In 1843, he was appointed colonial governor of New Zealand, an advancement that might have led to a knighthood and other important government posts. Instead, the experience ended as a brief and total political disaster. He landed in Auckland in December, in the middle of an ugly and protracted conflict between the Maoris and the white settlers over land rights. FitzRoy seems to have quickly offended all parties, not least of all the Home Secretary for the Colonies back in London, who was not kept informed of his actions. When word came on October 1, 1845, that Parliament had ordered that the new governor be recalled, an effigy of FitzRoy was hoisted by the settlers and joyously burned. Back in England, he returned to duty in the Royal Navy, where his reputation as an officer remained intact. He served as head of a dockyard and was assigned to put the navy's first screw-driven steamship, the *Arrogant,* through its sea trials. In 1850, he retired from active service, citing his wife's ill health and the need to tend to private affairs.

On the high seas, meanwhile, history and science were coming together. The California gold rush was under way, and vessels from around the world could not get there fast enough. Captains of naval and merchant vessels were beginning to take advantage of new navigational aides that were being developed as a result of new meteorological observations. A book by William Reid of the Royal Engineers, the American William Redfield's collaborator, offered storm-warning strategies, arguing that winds accompanying the movement of tropical storms could be predicted. At Reid's urging, weather observations were initiated at a number of British colonies and Royal Engineers stations overseas. The American naval officer Matthew Fontaine Maury, head of the Naval Observatory in Washington, had begun publishing *Charts of the Winds and Currents,* which was credited with dramatically shortening the time of many ocean passages. Efforts at collaboration led to the first International Conference on Maritime Meteorology at Brussels in 1853, which agreed to standardize the form and content of all weather observations at sea.

In Britain, the Meteorological Department of the Board of Trade was created in 1854 to implement the Brussels agreement with new instruments and logbooks aboard naval and merchant vessels. In Parliament, the suggestion by one member that such observations might one day lead to Londoners knowing "the condition of the weather 24 hours beforehand" was greeted by laughter. At the urging of the Royal Society, Robert FitzRoy was put in charge of the new agency and given the title of meteorological statist. A Royal Navy man seemed perfect for the job, and FitzRoy wanted it badly. The new office was to focus entirely on weather observations at sea, so what better man to bring Britain's naval and merchant vessels up to the new international standards?

FitzRoy plunged into the task, obtaining supplies of reliable instruments, writing instruction manuals, and organizing their distribution through the Admiralty to the Royal Navy's men-of-war and, through a network of port agents, to shipmasters of the mercantile marine. Then began the routine work of collecting and collating the data from their registers—valuable work in its way, of course, but not the kind of enterprise to engage a practical man of action and ambition. Meanwhile, he designed a sturdy shipboard barometer—known as the FitzRoy barometer—and wrote manuals for its use.

The new office translated from the German and published in 1858 an influential book by Heinrich Wilhelm Dove, a Polish scientist who was director of the Royal Prussian Meteorological Institute in Berlin and the leading European meteorologist of the day. Dove's *The Law of Storms* was the basis of much of FitzRoy's meteorological thinking. FitzRoy had always been more interested in science than most naval officers of his day, but unlike his American counterpart, Matthew Maury, he never pretended to be more than a follower of current thinking on the subject of storms and weather. Once he described himself as "only a superficial follower, however devoted an admirer, of real philosophers."

While Americans were trying to observe the pattern of winds around a storm as a whole, Dove was focusing on the change in the direction of winds that he observed as a storm passed a single location. His work represented a major advance in the science, showing a systematic pattern to weather changes, a predictability to the changes in the winds. Dove's theory found storms over Europe to be the result of collisions between a warm, moist equatorial air current and a cold, dry polar air current. His thinking anticipated by several decades methods of air-mass analysis developed by Norwegian scientists, and had FitzRoy's use of Dove's work been encouraged rather than suppressed, it might have led to important advances in theory sooner rather than later. Dove saw the rotating winds of middle latitude storms resulting from collisions between these air currents. Other features of a storm—such as air pressure, temperature, and humidity—were seen as consequences of these mechanical wind patterns. Whatever its flaws, FitzRoy's forecasting method represented a much more intelligent use of theory than would be employed by other forecasters for years to come.

Soon after the international conference in Brussels, a singular weather event that occurred during the Crimean War forced on European leaders the recognition of the military importance of meteorology and gave a boost to the idea of networks of weather observers. On November 14, 1854, British and French warships were wrecked by a surprise storm as they were anchored off Balaklava on the Crimea in the Black Sea. The French concluded that the vessels might well have been saved if the whereabouts of the storm had been reported to the fleet by way of the new telegraph as it advanced over the Continent.

In 1859, a great storm wrecked the modern ironclad vessel *Royal Charter* near the Isle of Anglesey off the coast of Wales with the loss of all aboard. FitzRoy investigated what became known as the Royal Charter Storm. Maps produced by his office showed a cyclonic storm moving over England in a tight formation. Its progress day by day was obvious, and the meaning of the advancing image was irresistible to his eye. The catastrophe could have been avoided if the storm had been reported in advance, he was certain. Now he knew the direction that the work of his Meteorological Department would take.

Soon after the Royal Charter Storm, the British Association for the Advancement of Science met in Aberdeen, Scotland, and called for the creation of a land network for the communication of weather conditions. Three weather districts over Great Britain and Ireland were defined. Reports from these districts would be posted in London at Lloyd's, the insurer, as well as other conspicuous places in the various "weather districts." Although the idea of predicting the weather was nowhere mentioned by the British scientists or by the Board of Trade, which approved the scheme, FitzRoy saw a great new opportunity. No longer was he in the business of merely collecting and tabulating data from the logbooks of ships at sea. He had been given a large and immediate challenge—something of consequence, finally, something worthy of his rank and vision. The veteran sea captain would do what shipmasters had always done. He would take his best measure of approaching weather and act accordingly. Now he was convinced that he had in his hands the means to warn more people of the dangers of approaching weather over a larger realm. There were lives to be saved. Now he was a man with a mission.

Like Joseph Henry a few years earlier in the United States, FitzRoy began establishing a network of telegraph clerks who were supplied with barometers and thermometers and were instructed to report daily weather conditions. By the end of 1860, he was receiving reports from 22 stations around the coasts of Britain and five daily reports from Europe. In February 1861, he issued his first storm-warning signals, employing a system of conical-shaped patterns of his own devising. In August, he began issuing daily weather forecasts, which were published in several London newspapers and drew widespread public attention.

Some leading members of the Royal Society were shocked and appalled by this turn of events. There were powerful reasons to avoid taking on an undertaking such as foretelling the weather, especially as a government-sanctioned enterprise in the middle of the nineteenth century. Still active in England were the so-called astro-meteorologists, who continued to profit handsomely by selling almanacs to the public. These almanacs trumpeted the old idea that responsibility for future weather, as for other future events, lay with the stars and the planets, the sun and the moon. Charlatanism was a real threat to legitimate science, and prophecy had long been its realm. The editor of *Zadkiel's Almanac,* himself a retired naval officer, and other astrologers had infiltrated the membership of the London Meteorological Society, the first London society devoted to the science; the organization had dissolved in 1842, its finances ruined and its morale shattered. Now here was another retired naval officer in an upstart government department making that old, dangerous, and disreputable claim: the ability to see into the future. Newspapers were publishing the forecasts, and editorials and commentaries weighed in on the subject. FitzRoy was gaining more fame and notoriety by the day, especially when his forecasts proved inaccurate. The popular press had great fun comparing the astrological prophecies of Zadkiel with the government forecasts of FitzRoy. To the conservative savants of the Royal Society, the threat to the reputation of science—to *their* reputations—was more palpable than any supposed public benefit from this untried and unwarranted weather forecasting service. What the new Meteorological Department was doing could only serve to elevate the status of astro-meteorology and degrade the public standing of true science.

FitzRoy was aware of these highly placed anxieties, although, perhaps unwisely, he did not pander to them. He was a confident and practical man who was providing other practical men with a tangible and valuable service. First and foremost, FitzRoy was a man of the Royal Navy, and a humanitarian at heart, and he saw himself on a mission to reduce the appalling loss of the lives of men at sea. He took pains always to distinguish his work from the flotsam of astrology by citing the science of Dove and by avoiding the idea that he was foretelling the future. With the storm warnings and weather prognostications, he was issuing a scientifically informed opinion, he said. Avoiding the term *prediction,* he referred to "cautionary warnings" and finally settled on a term of his own invention: he brought into the language the word *forecasting*. As he wrote, "Prophecies or predictions they are not; the term forecast is strictly applicable to such an *opinion* as is the result of a scientific combination and calculation, liable to be occasionally, though rarely, marred by an unexpected 'downrush' of southerly wind, or by a rapid electrical action not yet sufficiently indicated to our extremely limited sight and feeling. We shall know more and more by degrees."

In 1863, FitzRoy published *The Weather Book*, a meteorological guide he pointedly said was "intended for many, rather than for few, with an earnest hope of its utility in daily life. The means actually requisite to enable any person of fair abilities and average education to become practically 'weather-wise' are much more readily attainable than has been often supposed." This sense of popular utility, of science in the hands of everyman, offended some of the basic tenets of British men of science. To some members of the Royal Society and the British Association for the Advancement of Science, FitzRoy was playing into the hands of the seers and mesmerizers and frauds. Popular science was just another way of saying bad science.

While he nominally relied on the theory of conflicting air currents proposed by Dove, FitzRoy was hard-pressed to give it very rigorous scientific form while describing his forecasting method. He wrote in *The Weather Book* that he found it "extremely difficult to combine mathematical exactness with the results of experience obtained by practical ocular observation and much reflection." He sought to describe the intuitive sense that weather forecasters gain from experience and have always struggled to give satisfactory scientific form. The frequent telegraphic communications from widespread areas had provided him with "a means of feeling—indeed one may say mentally seeing—successive simultaneous states of the atmosphere over the greater extent of our islands." Terms like *feeling* and *mentally seeing* were not likely to impress men of reputation who were accustomed to practicing their science with the exactitude and objectivity of astronomy and physics.

Why did Robert FitzRoy kill himself? Could Darwin have been right about the organization of his brain needing mending? Were his fits of depression a genetic affliction? FitzRoy's maternal uncle, Lord Castlereagh, similarly had killed himself. He was the second *Beagle* captain to take his own life. FitzRoy had succeeded the command of Captain Pringle Stokes, who had shot himself in the cabin where FitzRoy had spent so many years. Certainly April 30, 1865, wasn't the first time that FitzRoy had driven himself over the edge, and certainly he was overwrought. He was ill, and he was losing his hearing. At the age of 59, he seemed to have aged rapidly. He was deeply in debt, never having been reimbursed by the Admiralty for substantial sums of his own funds he had laid out during the *Beagle* expeditions and the New Zealand mission.

No one took more personally the shattering scientific debate about the evolution of life that so occupied the second half of the nineteenth century. Darwin's publication of *The Origin of Species* in 1859 had deeply offended FitzRoy's conservative religious sensibilities. His close association with the undertaking may have inspired a feeling of personal guilt and public embarrassment. During a famous meeting of the British Association for the Advancement of Science in 1860, he had risen during a clamorous debate between Thomas Huxley and Bishop Samuel

Wilberforce about Darwin's theory of natural selection. The captain of the *Beagle,* at 55 looking older than his years, made plain that he regretted publication of *The Origin of Species* and rejected its contradiction of the first chapter of Genesis. In the last chapter of his own *Beagle* narrative, FitzRoy had expounded the view that the extinction of the mastodon and other large animals was due to their inability to fit through the doors of the Ark.

To his grieving wife, however, there was no doubt that it was the immediate problems of meteorology—the daily anxieties of public forecasting and the incessant carping of the scientific men, on one side, and the astro-meteorologists, on the other—that were weighing most heavily on her husband in the days before his suicide. The president of the Royal Geographical Society, Sir Roderick Murchison, voiced a widely held view that "if FitzRoy had not had thrown upon him the heavy and irritating responsibility for never being found at fault in any of his numerous forecasts of storms in our very changeful climate, his valuable life might have been preserved."

Under the best of circumstances, this sensitive, unstable perfectionist was temperamentally unsuited for such a job as forecasting weather. The mounting criticism of his mission and his methods from his peers seems finally to have pushed him over the edge. During his final days, he was deeply depressed. He had lost his famous confidence, his sea captain's ability to cope. Assistants had taken over his work at the Meteorological Department. The day before his death, he had met with the American Matthew Maury, a man more famous for his enemies than his friends, who had repudiated FitzRoy's weather forecasting techniques in articles recently published in Paris.

The controversy following the suicide of FitzRoy opened a divide in the science. Weather forecasting had given meteorology a public utility and a sudden popularity that the savants had never really expected, nor taken seriously. Before long, the pace and purpose of meteorology would change forever, although the official reaction to the death of Robert FitzRoy would ensure that England would not be leading the way into the future of the science. The Royal Society and the Board of Trade took the occasion of his death to investigate critically the performance of the Meteorology Department and the value of FitzRoy's storm warnings and weather forecasts. Their findings, released in 1866, might have been a foregone conclusion.

In 43 pages, their report demolished everything FitzRoy had accomplished. The 500,000 observations at sea that he had compiled were found to be inadequate for the purposes of scientific men. His forecasts were judged to be too inaccurate to be of use, although by modern standards they have since been found to have been in line with what could reasonably have been expected. In 1986, in the *Journal of the History of Science,* British historian Jim Burton wrote, "The report's figures were simply not fair by even the crudest assessments, but they were official and were quoted widely, the general view of FitzRoy's work suffering in consequence." The report concluded that the study of weather over the British Isles was better left in the hands of "a scientific society" rather than a government department. The Board of Trade responded by discontinuing the daily weather forecasts as well as the storm warnings, although public reaction to loss of the storm warnings was so great that they were restored in 1867.

The commercial men of the merchant marine, the maritime insurance industry, and the fishing fleet, among others, were much more forgiving of FitzRoy's inaccuracies than they were of the Royal Society's insistence on the everlasting purity of its science. They found an important ally in Charles Piazzi Smyth, the Astronomer Royal of Scotland, where the coast is famously vulnerable to North Atlantic storms. "High science is one thing, and storm warnings so completely another, that it is not fair to measure its use and right to existence by a test derived from anything else of so entirely different a nature," Smyth said in a letter to the Edinburgh Chamber of Commerce. The practical means to the prediction and warning of violent storms, he wrote, "are as abundantly in our hands now as they are likely to be for years, and even ages to come."

It would be another decade before routine daily weather forecasts would be resumed by the renamed Meteorological Office. In the intervening years, neither that agency nor the Royal Society accomplished anything notable to improve the accuracy or advance the science of weather forecasting.

11

Urbain J. J. Le Verrier

Clouds over Crimea

•

IN THE LONG HISTORY of European warfare, the Crimean War ranks among the biggest military blunders. In 1854, England, France, the Ottomon Turks, and Austria were allied against Russia, which was intent on extending its dominion over the Balkans and the Middle East. The Russians were fortified at Sevastopol, where for a year their Black Sea fleet was besieged by a flotilla of allied vessels. Among military brass on both sides, Crimea became a byword for strategic ineptitude and logistical incompetence. In Alfred Tennyson's "Charge of the Light Brigade," Queen Victoria's poet laureate glorified the ghastly annihilation of the British light cavalry in the battle for Balaklava: "Theirs not to reason why/Theirs but to do and die." The British nurse Florence Nightingale famously tended the wounded and the sick. Before it was over, about 250,000 soldiers and sailors had been lost by each side, more to disease than battle. It was a costly defeat for the czar and too costly a victory for England and France. Whatever else its consequences, however, the Crimean War encouraged the creation of national weather services in Europe.

During the evening of November 13, 1854, the sky darkened over the Black Sea and heavy rains began. Northeast winds swept over the fleet and the allied encampment, blowing stronger as the night wore on. By daybreak the tents and bedding were awash in a lashing southeast gale. Along the western and southern shores of the Crimean peninsula, winds continued to intensify, plowing into the fleet with the force of a hurricane, and carrying violent westerly squalls. And then came cold and snow. With the passing of the storm came the reckoning of enormous losses on land and sea. The British steamer *Prince* and its 7,000 tons of medical supplies, boots, and winter clothing was at the bottom of the sea. So was the 100-gun battleship *Henri IV,* the finest of the French fleet, and more than a dozen other vessels.

To an impatient public and press in England and France, the storm was just one more appalling debacle. And in the new age of telegraphic communications, new questions were being raised. It was clear to many people that the storm had battered western Europe a few days before reaching the Black Sea. Louis-Napoléon wanted to know if such disasters could be avoided, if a warning could have reached the hapless fleet in the Crimea that a great storm was on the way. Was the arrival of the storm at a particular time and place predictable? For the answer to this question he turned to one of the most famous scientists in the world.

Urbain J. J. Le Verrier had discovered the planet Neptune in 1846. It was a stunning scientific achievement. What especially impressed the public and press and the governments of the day was that he did it without any observations or instruments; instead, he made his discovery solely by the application of mathematical analysis, with a pencil his only tool. Studying the orbit of the planet Uranus, Le Verrier calculated that its eccentricity could not be explained by the gravitational influences of Jupiter and Saturn alone. He described a location in space and calculated a time when an unknown planet should be visible. He wrote to a fellow astronomer, Johann G. Galle at the Berlin Observatory, who trained his telescope on the spot and immediately saw an object that did not appear on any of the astronomical charts of the day. "The planet whose position you indicated *really exists,*" wrote Galle. On the strength of this discovery, Louis-Napoléon named Le Verrier to the directorship of the Paris Observatory after the death of François Arago in 1853.

In 1855, Le Verrier assigned an assistant, Emmanuel Liais, to study the Black Sea storm. He also appealed for observations for November 12 through November 16, 1854, from other investigators throughout Europe, including James Glaisher at the Royal Observatory in Greenwich, Christophorus Buys Ballot in Holland, Heinrich Dove in Prussia, and Adolphe Quételet in Belgium. From these observations, the scientists in Paris systematically traced the path of the cyclone across Europe and Asia. Le Verrier concluded that the arrival of the cyclone over the Crimea could have been foreseen, and that sailors could be warned of the approach of future storms. What was needed, he said, was the creation of "a vast meteorological network" across the continent. But Le Verrier would be frustrated for several years in his efforts to develop such a weather system in France.

Obtaining the cooperation of the various telegraph companies as well as the coordination of other national observatories for such a scheme naturally was more difficult in fractious Europe than in the United States. But Le Verrier's problem was more complicated. Among his fellow savants at the Académie des Sciences, conservative resistance to the idea of predicting future weather was deeply entrenched. Certainly Le Verrier was well aware of his mentor's uncompromising views about the progress of meteorology. Everyone remembered the admonition of the great Arago, Le Verrier's predecessor at the observatory, who in 1846 had been famously outspoken on the subject.

It so happened that some low-down Paris publisher had misappropriated the astronomer's good name and sold a book of "Arago lectures" that contained ersatz weather predictions. By 1846, the fraudulent book had gone into a fourth edition, and the eminent scientist was much aggrieved. It was this injustice that provoked him to write in the annual report of the Paris Observatory that year: "Never has word escaped my lips, either in private or in the course which I have delivered for upwards of thirty years; never has a line published with my consent, authorized any one to imagine it to be my opinion that it is possible, in the present state of our knowledge, to announce with any degree of certainty, what weather it will be a year, a month, a week, I shall even add, a single day, in advance." He added: "At present, I believe that I am in a condition to deduce from my investigations the important result which I now announce; *Whatever may be the progress of sciences, never will observers who are trustworthy, and careful of their reputation, venture to foretell the state of the weather.*"

Arago's meteorological investigations had been far-flung and influential. It was his studies that finally took the meteors out of meteorology, disproving the contentions of James P. Espy and others that they originated in Earth's atmosphere. Arago reclaimed meteors and shooting stars and comets and asteroids for the high science of astronomy. Putting this misconception to rest, he had looked carefully into the ancient idea that the moon and comets influence changes in the weather. Writing in the annual report of 1846, Arago declared: "These results clearly show, in my opinion, that the influences of both these bodies are almost insensible, and, therefore, that the prediction of the weather can never be a branch of astronomy, properly so called." While these were salutary scientific conclusions, to the reigning savants of the day, weather forecasting would have gained more respectability had it been retained as "a branch of astronomy, properly so called." This association with astronomy and comparison of the results of weather science to the divine exactitude of the heavens has haunted meteorology from the beginning. To the astronomers of nineteenth-century Europe, the empirical approximations of weather seemed hardly a science at all.

American meteorologists were most interested in tackling the subject of storms in order to contemplate their progress and predictability. This approach, pioneered by the German Heinrich Wilhelm Brandes, would come to be called *synoptic meteorology,* because it is based on simultaneous observations over a large area, and would be allied closely with weather forecasting. With the exception of Heinrich Dove, European scientists at midcentury were preoccupied with questions of theory, especially about the dynamics of the atmosphere. What causes rain and hail, lightning and thunder? A few international competitions offering handsome rewards were launched in the hope that scientists would be drawn to solutions to such problems, but the enticements attracted few takers.

The conclusion is hard to avoid that American scientists in the nineteenth century enjoyed a great advantage over their European counterparts by the absence in the young nation of large, old, powerfully entrenched philosophical organizations, the academies and societies royal, not only in the pursuit of weather forecasting but even in the fundamental investigations of the physics and chemistry of the atmosphere. The gatekeepers of European science wielded very heavy hands. In London, for instance, a single critical review by Henry, Lord Brougham had led Thomas Young to abandon his lectures, which anticipated the wave theory of light. In Paris, the savants of the Académie des Sciences were indisposed to accept even a significant breakthrough.

So it happened that Pierre Hermand Maille, an unknown amateur from the small provincial town of Saint-Florentin, was denied a place in the history of the science for his first description of the process of convection. Maille responded with a highly original paper to a competition sponsored by the Académie des Sciences in 1834 calling for scientific memoirs to explain the formation of hail. Not only did Maille explain the forces that cause hail, but—better than the American James P. Espy—he outlined the process of convection, condensation, and cloud formation. The Académie did not award a prize in the competition and did not publish Maille's paper. And yet, strangely, in 1841, the same savants couldn't seem to say enough about the importance of Espy's more loosely reasoned theory of storms. The physicist Jacques Babinet, an influential member of the Académie, conceded Maille's priority after the fact. Maille's original paper was not exhumed from the Académie's archives until 1965, and then at the request of British historian W. E. Knowles Middleton.

At midcentury, a special commission of the Académie was considering the installation of weather observation stations in the French colony of Algeria, and physicist Henri-Victor Regnault landed a telling blow at the flimsy theoretical base of the science. As Regnault put it, "We still do not know what must be observed, how it must be observed, or where we should observe it." At the same time, the aging physicist and astronomer Jean-Baptiste Biot argued that years of such observations in Russia had come to nothing, and "we now fall back on the hope of practical applications to justify the study of meteorology."

Only someone of Le Verrier's international stature, and only such circumstances as the Crimean debacle, could successfully overcome the institutional resistance of European scientists to the idea of forecasting weather. The advent of the telegraph made the military advantage of instantly communicating weather information increasingly obvious. Nothing has done more for the progress of meteorology than the exigencies of warfare. Still, it is a curious fact and a telling commentary on the state of the science in Europe that at the end of 1855, many years after American scientists Redfield and Espy had described the movement of storms, Le Verrier's report about the progressive path of the tempest over the Crimea was greeted in France as a revelation. The report described the movement of an "atmospheric wave" that could be read in the oscillations of the barometer and plotted in isobars on a map.

Still, progress came slowly. A Paris newspaper began publishing weather reports from France and other European stations in 1857, but the service envisioned by Le Verrier was years away. The first European storm-warning system was initiated not in France but in Holland, in 1860, by Christoph Buys Ballot, director of the Royal Netherlands Meteorological Institute, which he had founded in 1854. Admiral Robert FitzRoy began issuing such warnings through the British Board of Trade in 1861. The Paris Observatory began in 1863, although, as in Britain, the storm-warning service would be subject to considerable disruption and dispute.

The contrast between astronomy and meteorology could not have been more dramatic than in the personal experience of Urbain J. J. Le Verrier. Here was an astronomer who, using only his brain and his pencil, had been able to pinpoint the location of an unknown planet in the heavens. When he came to practicing meteorology, however, even with help of assistants, even with the resources of a national observatory, and even with a score of direct simultaneous observations from around the continent, the great Le Verrier was reduced to issuing only vague generalizations about the next day's weather. Outside the observatory, some called him a fool for trying, while others complained that he should do better. In 1863, Le Verrier was forced to sit ingloriously through an

Académie meeting while the leading savants of France listened to Louis-Napoléon's minister of war sharply question his claims of accuracy and criticize his forecasting methods. Even after the observatory began its own weather warning system, the Imperial Navy continued to rely on the bulletins it received from FitzRoy in England.

Inside the Paris Observatory, nothing was more disruptive to the service of meteorology than the temperament and character of Le Verrier himself. Autocratic and quick to take offense, Le Verrier was most ardently respected by those international colleagues who only occasionally spent much time in his presence. No one seems to have been able to work for the man very long. Several scientists left the observatory as soon as Le Verrier's appointment was announced. Emmanuel Liais, who had researched the 1854 Crimean storm, was fired by Le Verrier in 1857. Perhaps the most gifted meteorologist at the observatory, Deputy Director Edme Hippolyte Marié-Davy, who had launched its storm-warning service in August 1863, feuded with Le Verrier continuously. Finally, in October 1867, Le Verrier suppressed the forecasting service and devoted his energies to getting Marié-Davy fired. The deputy director appealed directly to the emperor, telling his side of the conflict and pointing out that since 1854, when Le Verrier was appointed, more than 100 people had left the observatory. After a mass resignation of the staff in 1870, Le Verrier was dismissed. The storm-warning service was resumed under Marié-Davy, but it was disrupted again in 1871 by the Franco-Prussian War, the abdication of Louis-Napoléon III, and the occupation of Paris by the communards. Le Verrier was reinstated in 1873, and the conflicts and crises resumed. At the important Vienna conference on international meteorology in August of that year, France was not even represented.

The most important French research of the period into the physics of the atmosphere came from an outsider, a mining engineer known as H. Peslin. In 1869, Peslin submitted to the Académie des Sciences a paper in which he had developed a mathematical equation for relating the spacing between *isobars*—lines of equal pressure—to the speed of winds. The paper was reviewed by an Académie commission, including Le Verrier, which decided not to publish it. The Peslin research was published in 1872 by Marié-Davy, who was by then directing another observatory at Montsouris. Five years later, it was translated into English by the American Cleveland Abbe and published by the Smithsonian Institution. At the Paris Observatory, meanwhile, there was no evidence that the Peslin formulation became part of the forecasting process.

12

Cleveland Abbe

"Ol' Probabilities"

•

"OLD PROBABILITIES" they called him—or, for short, "Old Probs." The nickname came when Abbe was just 30 years old, hiding his youth behind a full face of whiskers and thick, concave wire-rimmed spectacles that overcame his myopia. During his very first week of issuing weather "probabilities," the first official weather forecaster in the United States had commented that Cincinnati's half-inch of rain on Monday "was predicted on Saturday noon but is heavier than was expected." The handwritten notice was posted on the bulletin board of the Cincinnati Chamber of Commerce, with his simple error plain for all to see: "Weather Bulletin of the Cincinnati Observatory, Teusday September seventh 1869." Below the errant line, a meat packer had written, "A bad spell of weather for 'Old Probs.'" The term *probabilities* gave way to *indications* in 1876, and then, in 1889, Robert FitzRoy's coinage, *weather forecast,* became official; but despite such changes in nomenclature, Professor Cleveland Abbe always would be known, affectionately, as Old Probabilities. As Mark Twain, speaking in 1876 to the New England Society, would observe, "'Old Probabilities' has a mighty reputation for accurate prophecy, and thoroughly well deserves it." After his death in 1916, Abbe's son, Truman, looking through his father's diaries, would discover a quirky mental block for such a careful and learned man: Old Probabilities almost always misspelled *Tuesday.*

Like many a weather scientist, Cleveland Abbe's first love was astronomy. He had recently returned from what he always recalled as two wonderful years studying under the great astronomer Otto Struve in

Pulkovo, Russia, when, in 1868, the invitation to become director of the Cincinnati Astronomical Observatory came his way. This facility possessed one of the better telescopes in the country at the time, although it had been vacant 10 years and the property had fallen into disrepair. Abbe's personal interest in the weather in those days focused mainly on the need for astronomers to better understand and measure the defractive effects of various local atmospheric conditions. But Abbe was a young man of enthusiasm and ambition, and he knew an opportunity when he saw one.

He was born in 1838 in New York City, and during his boyhood the daily newspapers often had published articles by James Espy, Joseph Henry, William Redfield, Elias Loomis, and others describing the growing knowledge of the shape and character of American storms. Abbe would later say that the articles "convinced me that man should and must overcome our ignorance of the destructive winds and rains." While he was a 19-year-old graduate student at the New York Free Academy, Abbe recognized the possibilities of weather prediction from his reading of William Ferrel's classic paper of 1859 in Runkle's *Mathematical Monthly:* "I realized that he had overcome many of the hidden difficulties of theories of storms and winds; from that day he was my guide and authority."

After returning to the United States from Russia in 1867, Abbe found a job briefly as an aide at the Naval Observatory in Washington, where the Confederate Matthew Maury was long gone. From here he could see what had become of the pioneering telegraphic national weather observation system that had been organized by Joseph Henry beginning in 1849. The Civil War had wrecked the Smithsonian Institution's network, and a devastating fire in 1865 in the upper floors of the Smithsonian building had crippled its ability to rebuild the weather system. Henry had appealed unsuccessfully for the national government to take up the development of a new system. Meanwhile, there were those expanding lines of the telegraph, a proven tool for storm warnings, and here in Cincinnati was a scientific facility in a transportation and communications center, and a young scientist in search of a mission.

Abbe outlined these ambitions to the observatory's board of control in his inaugural report in June 1868: "If the director be sustained in the general endeavor to make the observatory useful, he would propose to extend the field of activity of the observatory so as to embrace, on the one hand, scientific astronomy, meteorology, and magneticism, and, on the other, the application of these sciences to geography and geodesy, to storm predictions, and to the wants of the citizen and the land surveyor."

To an old friend, meteorologist William J. Humphreys, writing years later, the scheme was typical of Abbe:

> [M]agnificent in scope and noble in purpose, but out of all possible proportion to a one-man observatory whose chief function hitherto had been that of entertaining the public. His disposition always was so hopeful that, apparently, he seldom took into consideration such obstructive factors as lack of time or want of opportunity. But if, perhaps, this accounts for his beginning some things that were never completed, it doubtless, on the other hand, also accounts for the completion of many things that otherwise might never have been begun.

Focusing now on meteorology, Abbe appealed to the Cincinnati Chamber of Commerce for financial support for a system of telegraphic reports and weather bulletins, which would be published in the local daily newspapers and posted at the chamber's offices. The information should be of great value to many business enterprises, he wrote in a letter to the president of the chamber in May 1869, especially agriculture and navigation. "As Cincinnati is very favorably situated with respect to the proposed outlying stations, it is most probable that 90 percent of our weather predictions will be verified. It is thus evident that we do not propose to guess at the weather, but (leaving that to the almanac makers) we shall be able to assert with confidence the nature of the weather for one, two, or four days in advance, as well as the stand of the water in the river." The chamber of commerce approved financing for the project for three months—not as long as Abbe had hoped, of course, but, as it happened, time enough to establish the value of the program and Abbe's credentials in the field of weather forecasting. Telegraph company clerks, Smithsonian volunteers, and others were employed. For the first time in the United States, in September 1869, a systematic daily forecast of the coming weather was in place. Abbe offered the daily service to the daily newspapers of New York, pointing out their value to shipping, and offered a daily telegram by way of the French cable to Urbain J. J. Le Verrier in Paris. In a letter to his father, George Abbe, in New York, an enthusiastic son boldly ventured a prediction: "I have started that which the country will not willingly let die."

Abbe was right. A critical political push came from Abbe's Smithsonian observer in Milwaukee. A civil engineer and geologist, Increase A. Lapham was a close student of meteorology who had started keeping weather observations in 1827. On the strength of studies by Espy and others showing that storms move from west to east, Lapham as early as

1850 had tried to persuade the Wisconsin legislature to finance a storm-warning system for the Great Lakes. In a January 1870 article published in a Chicago magazine, Lapham proposed that such a system be supported by the Chicago Academy of Science. The article illustrated the value of such a storm-warning system with a map that showed the progress of a storm over the region in 1859. The Academy of Science was supportive, but the *Chicago Tribune* ridiculed the idea. "It might be asked of what practical value such a department would prove if it takes 10 years to calculate the progress of a storm," it said.

But another newspaper, the *Milwaukee Sentinel,* had compiled a report of the numerous shipwrecks on the Great Lakes in recent years, noting the losses of life and property, and the long lists that were published in December 1869 proved especially persuasive to the cause of weather forecasting. Storms in the Great Lakes sank or damaged 1,164 vessels and killed 321 seamen in 1868, the newspaper reported, and the following year 1,914 vessels were damaged or destroyed and 209 people were lost. Commercial losses were in the millions of dollars.

Citing these losses, Lapham wrote his congressman, Representative Halbert E. Paine of Milwaukee, asking if it were not "the duty of the government to see whether anything can be done to prevent, at least, some portion of this sad loss in the future." Lapham drew up a petition asking Congress to support a system of storm warnings for the benefit of commerce on the Great Lakes. It so happened that Paine had been a student of the meteorologist Elias Loomis at Western Reserve College in Ohio at a time when Loomis was making his breakthrough studies on the structure and movement of storms. Were it not for this coincidence, weather services in the United States might have developed along very different lines. What Lapham envisioned was a regional system to be financed by "those most likely to be benefited." Paine took Lapham's petition and significantly enlarged its scope. Rather than calling for a regional system, Paine introduced a resolution in Congress calling for creation of a federally financed weather service for the benefit of the entire nation.

Abbe had serious misgivings about the fact that Paine proposed to put the responsibility for weather observations and storm warnings in the hands of the secretary of war and the U.S. Army's Signal Corps rather than an agency of civilians. Abbe had experienced firsthand the quality of weather observations made by military men. In a letter to Lapham on January 7, 1870, Abbe wrote: "The meteorological observations of the Army have generally proved themselves to be very unreliable and certainly no better than those the telegraph operators could easily make."

He thought it would have been wiser to form a committee of specialists to devise a plan of action. "And I am specially of the opinion that the money expended would do more toward effecting good results if it goes through the hands of meteorologists than through the hands of Army officers."

Abbe probably didn't know what powerful political forces he was up against at the time. Albert J. Myer, chief of the Signal Corps, was a commander without a command. Most Signal Corps servicemen had been mustered out with the volunteer army in 1865, leaving Myer with himself and an organization comprised of two lieutenants and two secretaries. "Immediately after the introduction of the measure, a gentleman called on me and introduced himself as Col. Albert Myer, Chief Signal Officer," Paine wrote. "He was greatly excited and expressed a most intense desire that the execution of the law might be intrusted to him." Paine wrote later that he gave the responsibility of a weather system to the secretary of war because it "seemed to me at the outset, military discipline would probably secure the greatest promptness, regularity, and accuracy in the required observations."

A New York medical doctor, Myer was especially well connected politically. In a dispute with Secretary of War Edwin Stanton during the Civil War, Myer had been fired as chief signal officer and waged a long but finally successful political campaign to get his job back. Along the way, he had become a friend of President Ulysses S. Grant and had cultivated numerous allies in the U.S. Senate. Taking charge of a new weather service was just one of many ideas that Myer was floating around Washington, and many politicians were interested in finding something for Myer to do. "Behind that resolution was clearly visible the hand of Myer himself," wrote a Signal Corps historian, "grasping for something that would keep the Signal Corps alive."

More generally, a larger political imperative argued for a beneficial national purpose for the North's victorious army. With the country still feeling the wounds of its terrible, fractious war, the politics of Reconstruction welcomed investment in a new national public service intended to help citizens cope with a powerful, nonsectarian peacetime enemy: America's violent weather. The legislation enjoyed widespread support. Within two months of its introduction, Public Resolution No. 9 was approved by Congress and, on February 9, 1870, it was signed into law by President Grant.

Abbe's misgivings about military control were quick to materialize. As would become increasingly evident, however many civilian lives and livelihoods might be at risk, the U.S. Army did not consider the weather a very dangerous enemy. It took three weeks for Secretary of War William W. Belknap to officially inform Myer that Congress had given his Signal Corps a new mission. At the end of June, Myer had no forecasters on the payroll and apparently no forecasts in his plans. He reported that the Signal Corps had limited its activities to "practically testing the promptness with which reports will be received and the facts as to the approach and force of storms which synchronous reports following each other in close succession will announce, without any effort of anticipation. It has been considered wise by this Office not to attempt any more than this at the outset." Modeled after Abbe's operation at the Cincinnati Observatory, weather observations from various military posts would be telegraphed daily to Washington, but the Signal Corps brass would go several more months without issuing any storm warnings or weather forecasts.

Pressured by Lapham and Paine, Myer finally met Lapham for the first time on November 8, 1870, as the violent storm season was kicking up around the Great Lakes. He quickly hired Lapham as the special civilian assistant to the chief signal officer. Myer put Lapham in charge of weather forecasting for the Great Lakes and ordered all weather reports rerouted to Chicago. Within minutes of his appointment, Lapham issued this first official U.S. government storm forecast: "To all observers along the Lakes: Bulletin this at once. A high wind all day yesterday at Cheyenne and Omaha. A very high wind reported this morning at Omaha. Barometer falling with high wind at Chicago and Milwaukee. Barometers falling and thermometers rising at Detroit, Toledo, Cleveland, Buffalo and Rochester. High winds probable along the Lakes."

But Lapham would not last long. He was 60 years old, and his health was not good. It would take a younger man to cope with the frustrations of the job. Soon Lapham returned to his own business and professional interests in Milwaukee.

Finally Myer turned to the only man in the country with firsthand experience forecasting weather. For Cleveland Abbe, it was a turning point. As recently as December 1869, even while advocating expanded weather services in Cincinnati and elsewhere, Abbe harbored hopes of returning to the science that first attracted him. So he wrote in a letter to Lapham on December 7, bringing his observer up to date about progress in Cincinnati: "Negotiations are now in progress which will doubtless lead to the call of an able meteorologist to take charge of this matter and the attempt to forewarn important ports of approaching storms will be inaugurated and pushed systematically. I hope that I shall thus in a few months be able to return to my proper study—astronomy—and to extend the labors of the observatory in that direction." But his Cincinnati plans did not go well. In the polluted air of the industrial city, the observatory building was run down and unsuitable for astronomical observation. Moving it to a new location was costly far beyond the interests of its private sponsors in the science. They soon decided to divert their funds to the founding of the University of Ohio.

Cleveland Abbe began work for the weather service of the U.S. Army Signal Corps on January 3, 1871. He was given a civilian scientific position with the title of professor, in line with tradition of military academies. For this work, the astronomer-turned-meteorologist relied heavily on the studies of James P. Espy on the movement of storms and the theories of William Ferrel about the circulation of the atmosphere. "After a month's practice, it was decided that my forecast would evidently more than fill the popular expectations and tri-daily forecasts began at once," Abbe wrote. Although he expected to resign in three years, at the end of the term of his appointment, Old Probabilities would stay at it for more than 45 years.

Despite the civilian and scientific nature of its mission, the Signal Corps' Division of Telegrams and Reports for the Benefit of Commerce was driven throughout by the military character of its organization. Rules and regulations for the many enlisted men were strict and according to code.

At the same time, the attitude of the few officers toward the scientific nature of their work was casual, at best. A memoir on the use of homing pigeons for carrying messages was written, and an extensive compilation of weather proverbs from across the country was produced. Forecasting was accomplished entirely by rote, connecting observations of like values of barometric pressure by drawing isobaric lines, which identified areas of low pressure, which were assumed to be centers of storms. That was as far as it went, and just about as far as it would go for many years in the United States.

Myer had no interest in scientific research. Throughout his time as head of the weather service, he employed but one research scientist. The natural scientific advantage that earlier investigators had seen in the continental scope of the American landscape was squandered during these years. If new insights were going to emerge about the motions of the atmosphere or the structure of storms, it was not going to happen in the United States. Weather forecasting may have been the mission, but there was no doubt that the War Department was the master. Observer recruits received meteorological instruction and other training at an old fort across the Potomac River in Arlington, Virginia. They were instructed by some of the best weather scientists in the country, including Abbe, Ferrel, and Loomis. Like servicemen everywhere, however, what most of them seemed to recall of their Signal Corps training was not the weather science but the bedbugs, the bad food, and the rigid boot camp culture of the place.

Myer died in 1880 and was succeeded by General William B. Hazen, who was much more interested in supporting meteorological research that Abbe argued was critical to the weather service. But the Signal Corps was wracked by internal dissension and an embezzlement scandal, and, as an economy measure, meteorological instruction of recruits was discontinued. The end of military control of the nation's weather service was in sight. Meanwhile, for scientific men, their service in the Signal Corps was bitterly frustrating. Without the nurturing presence of Cleveland Abbe, in all probability, most of the eminent civilian scientists like Ferrel, Loomis, and physicist Thomas C. Mendenhall would not have come to work in what became known as the Study Room and laboratory. And certainly they would not have stayed as long as they did.

From the outset, Mendenhall wrote later, Abbe realized "that the state of our knowledge of meteorology then was quite inadequate for anything like accurate forecasting" and consistently sought funds from the War Department for a systematic study of the subject. From the start, "Abbe's plans met with obstruction at almost every turn."

At its headquarters near the War Department in Washington, the weather agency was literally and figuratively split in two. On one side of G Street NW were the researchers in the Study Room, in the physics laboratory, and the Instrument Testing Room. On the other side of the street were the army brass—the chief signal officer, his military aides, the property and disbursing officers, and the operations forecasters. Mendenhall described the frustration of scientists who were met with perfunctory military orders to conduct research into this or that unknown phenomenon and with the interminable red tape of a disinterested military establishment.

Mendenhall recalled "a sort of a tradition among military men . . . implying that a properly signed written order from a superior officer to do a certain thing carried with it not only the duty of doing it but also the capacity to do it, which I imagine may be a very stimulating idea for one engaged in battle though of doubtful value in scientific research." Scientists fumed about the military style, and officers bridled at their resistance. "Unfortunately Nature does not yield her secrets in response to orders," Mendenhall wrote, "and there were naturally many failures to 'get results' on time."

The ham-handed military culture only exaggerated the elements of sharp division within meteorology that had emerged during the second half of the nineteenth century. On one side, the practical men representing commercial interests, public safety, and convenience saw meteorology as a service. On the other, scientific men saw in the motions of the atmosphere great mysteries that could only be solved in fundamental research. In England, where matters were in the hands of the scientific elite of the Royal Society, Robert FitzRoy's weather service had lost. In the United States, in the hands of the U.S. Army Signal Service, weather science went begging. At the time, few men anywhere were in a position to bridge the growing divide. One of them was Cleveland Abbe.

Old Probabilities was neither a great research scientist nor a great forecaster. "He made no important discovery, and published but little that contained anything new and original," wrote Humphreys. "He was primarily a teacher and a propagandist. He compiled in form convenient for use the best that was known about meteorological instruments, about climate and crops, and about the mechanics of earth's atmosphere; published nearly 300 papers on meteorology and kindred subjects; and wrote many thousands of letters encouraging others to contribute something to our knowledge of the atmosphere and its phenomena." He may have been the first in the nation, but Abbe was not even an outstanding weather forecaster. "His success in this role was indifferent," recalled Humphreys. "He was not one of the few that see a weather map in its entirety and quickly perceive its probable changes during the next twelve to thirty-six hours."

For nearly half a century, as the chief scientist of an agency hobbled first by cumbersome peacetime military dictates and later by political infighting within the government bureaucracy, Cleveland Abbe's greatest skills were human. He was singularly unselfish, giving time and praise liberally to his associates. He was deeply respected, even beloved. "His disposition was most amicable; he was one of the most lovable men that I have ever known," wrote Mendenhall. "There was a rare simplicity and frankness in his speech which was reflected in his acts, and added much to the charm of his personality."

In 1891, the nation had had enough of its experiment with a military weather service. Congress approved legislation transferring the weather service to the Agriculture Department. The agency then became known as the Weather Bureau.

13

John P. Finley

Down Tornado Alley

•

AS THE WESTERN EXPANSION populated larger tracts of the prairie and the Great Plains, a diabolical surprise confronted the nation. More and more settlers found themselves in the paths of the most horrifying storms. The violent tornadoes were not unique to this part of the country, of course, but there was definitely more of them than anyone had been led to expect. It was a shock, and maybe even a threat to the growth of the nation. Where were these terrifying storms coming from, and why? Nobody really knew. There were suggestions that the clearing of land or the railroad tracks or the electrical lines of the new telegraph or some other agency of civilization was to blame.

The Signal Corps routinely assigned an observer to investigate the aftermath of an outbreak of tornadoes. In the beginning, details of the tornadoes' tracks and destruction were duly recorded and, for the most part, filed away. William Ferrel laid down the theory of the storm, explaining how its great centrifugal force accounts for the low pressure inside the whirl. But much remained unknown about the conditions that bred such things. Like the immigrants of the prairie and the plains, the nation's new weather service wanted answers to questions of a practical bent: Could tornadoes be predicted?

The study of the problem changed in character and scope in 1879 with the assignment of a particular young, uncommonly dedicated, college-educated recruit to study the tornadoes that ravaged Kansas, Nebraska, Missouri, and Iowa on May 29 and May 30 of that year. The research conducted by John Park Finley was some of the most careful and comprehensive weather science of the era. It earned him an international reputation and placed him far ahead of everyone else in understanding this phenomenon that so peculiarly haunts the middle of the American continent.

Maybe Finley was too far ahead of his time. A nation eager for the enrichments of Manifest Destiny was not really so eager to hear Finley's dire warnings about the terrifying storms out there on the prairie and the plains. Certainly the officials of a new national weather service authorized expressly for "the benefit of commerce" were not enthusiastic. Finley's study found many more tornadoes than anyone in Washington had expected or evidently wanted to hear about. The brass of the Signal Corps rejected his recommendations, eventually abolished funding for his research, and even forbade the use of the word *tornado* in any weather forecasts.

The tornado research of Finley marked the culmination of the general definition of American storms that had begun early in the nineteenth century. Words such as *tornado* and *hurricane,* which had been bandied about interchangeably at the beginning of the century, by now had fairly specific jobs to do. The work by William Redfield and William Reid had identified the shape and character of tropical storms and hurricanes. The investigations of Elias Loomis and the theories of James Espy and Ferrel had fairly successfully distinguished the character of midlatitude cyclones, the big seasonal storms that cross the continent. In 1879, Finley argued for technical discipline of nomenclature among researchers. "It is not that we lack the coinage of suitable terms, but that having them we fail to insist upon their absolute use for the undeniable advantage of clearness and rapidity in investigation," he wrote. "It cannot be denied that the class of storms which I have been ordered to investigate are tornadoes, and nothing else, their unfailing accompaniments being quite fully exposed in this report, as well as in others of earlier date."

By the end of the century, at least the experts had sorted out the differences. In 1899, Weather Bureau chief Willis Luther Moore, writing in *National Geographic Magazine,* chided "the press, and nine out of ten people who should know better" for not correctly making the distinction between a cyclone and a tornado. "The cyclone is a horizontally revolving disk of air, covering the whole United States . . . with the air-currents from all points flowing spirally inward toward the center, while the tornado is a revolving mass of air of only five hundred to one thousand yards in diameter, and is simply an incident of the cyclone." Considering his agency's ambivalent approach to the subject, Moore might have forgiven the public a certain confusion. In the tornado capital of the world, the nation's weather agency would conduct virtually no research into this type of storm for the better part of half a century, and the forbidden word would not appear in an official public weather forecast until 1952.

Whatever the state of the science, clearly more than meteorology was at stake. Thompson B. Maury, who in 1871 had become one of the first civilian forecasters for the Signal Service, complained in 1882 about

> a widespread impression that, with the deforesting and settlement of the West, tornado-visitations have increased, so that a prominent journal recently raised the question whether their frequency and destructiveness will not have "a permanent effect on the settlement and prosperity of the country." We are even told that in some places the alarm created by these storms is so great that "the people are not only digging holes in the ground and building various cyclone proof retreats, but in many instances persons are preparing to emigrate and abandon the country entirely."

Although Maury had since resigned from the Signal Service to write about the weather for the *New York Herald,* his article in the September 1882 edition of the *North American Review* certainly reflected official thinking on the subject.

While settlement might "slightly modify local climatic conditions," Maury wrote, "man is as powerless to work any change which will augment or diminish the number of tornadoes, or to disturb the ponderous atmospheric machinery which produces them, as the puny fly is to retard or accelerate the motion of a powerful steam-engine." And while the nation's weather service was in a position to warn inhabitants of conditions that were ripe for severe local storms, "the successful prediction of a full-fledged tornado is a triumph yet to be won by meteorology." Moreover, this was not a direction the nation's weather service was really planning to go. "It is probable that, for local warnings of this kind, each community will always have to rely mainly on itself, or upon its own state weather service."

John P. Finley and his superiors had completely different opinions about how the weather service should handle the problem of tornadoes. At the policy remove, national officials seemed fatalistic and passive, worried more about false alarms and panic and unnecessary disruptions of business. They seemed intent on assuring citizens of the minimal risks tornadoes posed. Finley saw the problem from a much closer range and took exactly the opposite approach. It was a question of life and death. He was aggressive and outspoken, choosing his words like a man trying to sound an alarm. "The populous region of the United States is forever doomed to the devastation of the tornado," Finley wrote in 1887. "As certain as that night follows day is the coming of the funnel-shaped cloud. So long as the sun shines upon the vast regions in the Mississippi and Missouri valleys, there will forever occur those atmospheric conditions which terminate in the destructive violence of the tornado."

Finley seemed to be have been deeply affected by the casualties and the psychological aftermath of the storms he investigated so extensively in 1879 across the central plains. He noted in his report that the 42 deaths and the 262 houses destroyed that May were not large losses when compared to some tornadoes that had ravaged the populous eastern states. He wrote:

> It must be remembered that these storms, especially in Kansas, passed over very new and thinly settled districts. Even this loss is a terrible calamity to befall young and feeble communities struggling for a foothold in the vast prairie wilderness.
>
> The effect upon the people was pitiful in the extreme. Night after night (particularly in Kansas) hundreds of people never went to bed, but remained dressed and with their lanterns trimmed, watching for a fresh onslaught which they expected momentarily. Every dark cloud or sudden increase in the velocity of the wind seemed to them filled with evil forebodings, which could not be allayed until every vestige of supposed danger had vanished. The terror depicted upon the countenances of the bravest men, at the sight of a dark cloud above the horizon, was something beyond description or realization, except by those who could witness their excitement. Persons were preparing to quit the country; business of every kind succumbed for a season, except that of generously supplying the wants of the sufferers by well-organized relief committees. Many acts of devotion and self-sacrifice redound to the glory and honor of Kansas people.

These were the people and these were the days that so moved John Finley to spend the next 10 years researching tornadoes and advocating ways to warn citizens away from their paths. "It was a matter of the most serious consideration, and of oft-repeated inquiry, whether or not this region was particularly subject to this class of storms, and the frequency with which they might and would occur," he wrote. "Could the question be settled? Would it be settled by the Signal Service Bureau? Does the Signal Service Bureau pay particular attention to this class of storms, so intimately affecting our welfare? Will the bureau be able to forewarn us next spring and summer? These and similar questions were earnestly propounded by the active business men of every community, both within and outside of the storm's path."

John Park Finley was among the class of college-educated new recruits that Cleveland Abbe had tried to attract into the ranks of the Signal Corps' weather bureau. Born in 1854 in Ann Arbor, Michigan, the son of a successful farmer in Ypsilanti, Finley was uncommonly educated for his time. After attending primary schools in Ypsilanti, he completed a course in classical studies at the State Normal College, obtained a bachelor of science degree and later a master of science at Michigan State Agricultural and Mechanical College, where he studied the effect of weather and climate on agriculture. He spent another year studying law at the University of Michigan. Even after enlisting in the Signal Corps in 1877 and completing its meteorological instruction, Finley enrolled as a graduate student at Johns Hopkins University in Maryland with the intention of further preparing for his research into tornadoes and cyclones.

Finley was well connected politically, not only through the local prominence of his father, Florus Samuel Finley, but nationally through his uncle, Hiram Berdan, who lived in Washington. As a colonel during the Civil War, Berdan had formed a famous unit, Berdan's Sharpshooters, for the Army of the Potomac, and was promoted to the rank of general at Bull Run. This connection brought Finley to the top of the list of recruits and proved critically helpful during several controversies in his Signal Corps career.

A big bear of a man, six-foot-three and well over 200 pounds, Finley was widely respected for his scientific acumen and uncommon dedication. But he was not personally admired by his superiors or subordinates. He was very ambitious but apparently not very astute politically. While Berdan's connections helped him rise through the military ranks, he seems to have lacked the personal skills or administrative talents of a successful officer. And while his uncle's pull with the army brass was great enough to get him out of trouble, Finley seemed unable or unwilling to stay out of it. Tornado forecasting was just one of his battles with the brass. He worked too hard, for one thing, putting in relentlessly long hours and expecting everyone around him to sacrifice as much. In 1882, less than a month after his marriage to Julia V. Larkin, he was pushing himself so hard that he ended up in an army hospital in Washington on the verge of a nervous breakdown. An army physician diagnosed "premonitory symptoms of neurasthenia" from "too much brain work" and prescribed much-needed rest. "Should he continue to do as much as at present the consequences will be of very grave character," wrote the doctor.

Finley was first to show that tornadoes associated with midlatitude cyclones are confined to the southeast quadrant of the storm. He was first to explain their southwest-to-northeast paths as part of the upper flow of cold northwest winds behind the squall line. He was first to identify the set of surface conditions that promote the formation of tornadoes. His tornado chart for the first time identified the intrusion of a dry line of air into the warm moisture of the south winds as critical to tornado formation. Describing Finley's work in *Science* magazine in 1884, William Morris Davis noted optimistically: "The limitation of tornadoes to certain parts of cyclones . . . is a most hopeful sign, that, with longer and more detailed study, the smaller storms may, in a few years hence, be predicted with as much accuracy as the larger ones are now."

Finley organized a network of tornado reporters that numbered more than 2,400 people at its peak. He experimented with tornado predictions, although his claims of success were open to debate. In the next issue of *Science,* he reported that he had successfully predicted on 55 days in the spring of 1884 that no tornadoes would occur; and that on the 28 days tornadoes were predicted for particular states or larger regions, they occurred in the region where they were predicted on 17 of those days, and on 11 days they occurred where they were not predicted. Mathematicians scolded Finley for lumping the common events of days with no tornadoes together with the rare events of tornadoes. Finley argued that on some days it was just as important to know, and as difficult to determine, that no tornadoes were expected.

His claims and arguments did not persuade the Signal Corps. His recommendation that a special observer be stationed in Kansas City during the tornado season to help warn citizens of the plains was not pursued. A ban on forecasting tornadoes was imposed in 1885, lifted briefly in 1886, and reinstated early in 1887. The chief signal officer decreed that "neither the present condition of the science of meteorology nor the practical needs of the country would justify such forecasts" and that "more harm would be done by the prediction of a tornado than from the tornado itself."

Cleveland Abbe would echo that sentiment in 1899, after tornadoes raked across Missouri and Iowa in April. The *Chicago Tribune* wondered whether "in these days of telephones and telegraphs" it was not possible to warn residents of towns in the path of tornadoes. The answer, said Abbe, was no. "It is certain that if any such arrangement were possible, the Weather Bureau would have done this many years ago," he wrote, "but the time has not yet come." The best it could do was what it had done since 1871: to warn that "severe local storms are probable" in a region. The word *tornado* was still forbidden. "We have no right to issue numerous erroneous alarms," Abbe wrote. "The stoppage of business and the unnecessary fright would in its summation during a year be worse than the storms themselves, so few and so small are they."

Eventually, funding for Finley's research dried up, and he was assigned other duties. For several reasons, the time to take advantage of his science was not right. It would be several decades before scientists realized the real value of Finley's research and how much, in other circumstances, he and others might have accomplished. Part of the problem was the character and personality of the nation's first severe storms forecaster. Tornado forecasting could have had a better advocate. When young recruits petitioned their superiors to protest abusively harsh treatment during training, Finley was one of four officers who were singled out. In 1886, during a special congressional investigation that eventually would lead to the transfer of the weather service to the Department of Agriculture, there was Finley in the middle of it, offending his superiors with his testimony.

In April 1886, when press reports were describing the desperate circumstances of six Signal Corps men marooned in the Mountain Meteorological Station atop Pikes Peak in the Colorado Rockies, Finley's bosses must have figured they had the perfect assignment for this troublesome character. Fierce storms and snowslides had wiped out telegraph lines and trails, and the summit station had not been heard from in several months. "None of the hardy mountaineers could be hired to make the attempt," recalled Finley, because April was considered the most dangerous time of year for an ascent. And the Signal Corps observer at the base station in Colorado Springs "was anxiously hoping that I could be induced not to venture the trip to the Summit, as I had informed him that he must accompany me."

Of course, nothing was going to dissuade Finley. "We were completely sewed up in several layers of gunny sacks, provided with alpine pikes, hunting knives and small revolvers, as well as the instruments I was to use for comparison work at the summit," he recalled. "The mercurial barometer was fastened to my back, after being covered in its wooden case with gunny sacking; the thermometers, likewise protected, were carried by the observer." At the crest of the summit, in the middle of a heavy snowfall, Finley lost his footing.

> The observer staggered on a few yards to the station, and fell against the door more dead than alive. The men within rushed to the door to respond to the knock that seemed to come from another world. They

> had given up all hope of rescue. The observer barely had sufficient strength to explain about me and that I must have fallen at the crest. All hands rushed out and finally found me in the blinding snow by the upright position of my alpine pike. I was completely covered with snow and looked like the rocks all about me. I was carried to the station and when I regained consciousness, found myself on a big office table, prostrate with men working over me to restore circulation.

Finley's position in the Signal Corps may have been difficult even for a man with more political and personal talents. The agency was becoming more and more deeply fractured by infighting between military operations officers and civilian scientists. To the line officers of the army, Finley was a college-educated junior officer who had come up too quickly and too easily through the Signal Corps ranks. To the civilian scientists, he was a take-no-prisoners military man. When the end to military control of the weather service came in 1891, Finley had the same choice as everyone in the Signal Corps. He could stay in the U.S. Army or he could become a civilian employee of the Agriculture Department. He would remain interested in meteorology all of his life, but the man who had worked so hard to achieve his military rank chose the Ninth Infantry.

14

Mark W. Harrington

Civilian Casualty

•

THE FIRST CIVILIAN chief of the new U.S. Weather Bureau was a scientist with a national reputation for excellence who was in a good position to take the agency to important new heights in the study of meteorology. After 20 years of military control, there was reason to suppose that the advent of civilian authority would mean an important shift in the approach to the weather agency's pursuit of its young science. The *New York Tribune* advised President Benjamin Harrison to select a competent scientist and to ensure that the position was not political. "The general spirit which pervades the administration of this bureau at the outset will probably be perpetuated for years," it said. In every respect, professor Mark Walrod Harrington of the University of Michigan, astronomer and meteorologist, was eminently qualified for the position of leadership in the new agency of the Department of Agriculture. By all accounts in 1891, Secretary of Agriculture Jeremiah M. Rusk had made an excellent choice. It so happened, however, that the U.S. Weather Bureau was not on its way to becoming a leading scientific organization, whatever the needs of the science. Events and succeeding politicians conspired against the nation's new weather agency and against the new chief, and within a few years he was driven from office. Whether this experience of being fired by the president while at the top of his career was what drove him so completely insane is hard to know.

Born in 1848 in Sycamore, Illinois, the son of a physician, Harrington was raised on a farm. He attended Northwestern University until his sophomore year and then enrolled at the University of Michigan, where he earned a bachelor's degree in 1868 and a master's in 1871. That summer he traveled to Alaska to serve as an astronomer's assistant for the U.S. Coast and Geodetic Survey. Harrington returned to the University of Michigan, where he had begun teaching before graduation, and became an assistant curator at the museum and an instructor in mathematics and several sciences. He studied at the University of Leipzig from 1876 to 1877, then traveled to China, where he served as a professor of astronomy in the cadet school of the Foreign Office in Peking. In China, Harrington became ill, forcing his return to the United States, where in 1878 he joined the faculty of Louisiana State University. The following year he returned to the University of Michigan as a professor of astronomy and the director of the institution's observatory in Detroit, where he taught meteorology as well as astronomy. In 1884, he founded the influential *American Meteorological Journal* and served as its editor until 1892. Thus, in 1891, the secretary of agriculture selected as Weather Bureau chief a worldly scientist who, at the age of 42, was uncommonly acquainted with the state of meteorology and its researchers.

Harrington brought to the job a new vigor and a new intellectual style. Under the Signal Corps, no one outside of Washington and none but a select few were allowed to forecast weather. Harrington took over a national weather forecasting agency that had only four trained forecasters. In three years, he enlarged the number of trained forecasters to 40 individuals and scattered them around the country. The national forecasts were issued as before, twice daily for up to 36 hours in advance, but forecasters in the new district offices were welcome to amend them to reflect local needs and conditions.

As expected, Harrington for the first time set down for the Weather Bureau the foundation of a meteorological research program, much of it focused on the needs of the nation's farmers. He initiated climatological studies, collected rainfall and snowfall records in a useful new way, and launched a cooperative project with the University of Michigan and Johns Hopkins University that studied the physical properties of soils. The extension of weather services from navigation to agriculture, he wrote, presented forecasters with "several new problems and rendered some of the old ones more complex." While navigators were interested mainly in episodes of high winds and the arrival of large, damaging general storms, farmers also wanted to know the timing and extent of rainfall and the onset of the small, local storms that irrigate crops during the growing season as well as more detailed temperature and humidity conditions. Meeting all these needs, and the demands of Congress, Harrington said, was "a very difficult task, fairly impossible in the present state of our knowledge." Harrington had a clear understanding of the state of the art of weather forecasting, the limitations of its current methods, and undoubtedly would have liked to embark on fundamental new studies of the atmosphere.

In the journal *The Review of Reviews,* Harrington made the case for such fundamental research in an 1895 article titled "The Value of Weather Forecasts to Agriculture and Inland Commerce." Harrington said the weather map remained an important tool in forecasting,

> but as a means of adding to our knowledge and forming a true study of atmospheric changes its usefulness is almost exhausted. Further advance is necessary if we are to successfully forecast local storms, fogs, rain, snow, hail, moderate temperature changes and the like, but this advance is probable only by the efforts of competent scientific students, chiefly physicists and mathematicians. There must be the opportunity for some Galileo, Kepler, Copernicus and Newton if we are to lift the art of weather forecasting from its present ptolemaic stage into the stage of true theory as they lifted astronomy.

Only government was equipped to undertake such a study because meteorology, like geology, was "a general terrestrial science" beyond the scale of private resources, Harrington argued. Nevertheless, the prospective benefits were great. He offered no guarantees of scientific results, but he certainly was optimistic. "Some of the improvements, the forecast of fog, for instance, would come very easily," he wrote, probably in a year of systematic investigation by a competent scientist. "Others, as the true and complete theory of local storms, might not be completed for a century. I have estimated that three competent physicists, left to pursue their investigations for ten years without disquiet and given proper encouragement and assistance, would probably be able to so improve our art of weather forecasting as to satisfy all ordinary requirements."

Harrington would not get the opportunity to test these ideas, or even to make more than a beginning in the way of meteorological research at the U.S. Weather Bureau. The economic conditions were not right. The country was heading into a severe recession, and federal funding was shrinking. More important, the political conditions in Washington were wrong—all wrong—for the progress of meteorology in America. If ever there was going to be "ten years without disquiet" in the Weather Bureau, Mark Harrington was not going to see them.

It so happened that Harrington's appointment had come during the Republican administration of Benjamin Harrison, who in 1888 had won the vote in the Electoral College but had lost the popular vote to incumbent Democrat Grover Cleveland. The stinging political anomaly opened a rabidly partisan divide in the capital. Although recent civil service reforms were supposed to protect most federal employees from the old political spoils system, the election of Cleveland to a second term in 1892 put all Harrison appointees at risk. Cleveland's appointment of Julius Sterling Morton of Nebraska as secretary of agriculture meant trouble for every government scientist. Like most politicians, the new man in charge of the Department of Agriculture was immaculate of science and distrustful of its claims to independent authority. Turmoil followed in all government agencies with scientific missions: the Weather Bureau, the Geological Survey, the Coast Survey, and the agricultural experiment stations.

Morton came to town in February 1893 and immediately began consulting not with Chief Harrington, but with Major H. H. C. Dunwoody, the chief forecaster under the Signal Corps. He was the officer who had compiled the book of weather proverbs from around the country and who throughout the 1880s had vigorously opposed efforts to expand scientific research into the subject. Dunwoody and Morton were of the same mind about the value of scientific research and government support of it. And as soon as Morton took over, the old political game was on. Summarily, Morton appointed several of his own "experts" to the staff of the Weather Bureau. Harrington dismissed them, calling them "fiat experts," and observed that "friction always occurs when an attempt is made to turn a Bureau like mine into a refuge for politicians." By April, Harrington found himself the subject of an investigation by the attorney general's office into charges of a variety of misdeeds. While Harrington understood the political nature of the conflict with Morton, he may have overestimated the moral character of his opponent.

Morton fired Harrington's assistant chief and put Dunwoody in his place. He chose as the government investigator an assistant attorney general who was himself under investigation for malfeasance. Predictably, in late May, the prosecutor issued a damning report. Refusing to yield, Harrington appealed directly to President Cleveland, calling the report "a willful and malicious falsification of the testimony taken in the case." The *New York Tribune* called the report "a profound surprise" to "everyone who had followed the testimony taken during the investigation." If Cleveland were to "take the trouble to look into the matter" he would find that the investigator "has made exactly such a report as a man would make who had received orders to convict and had obeyed them to the letter." In his letter to the president, Harrington appealed to Cleveland for a fair hearing, telling him that it "is useless for me to demand hearing from the Secretary of Agriculture." Evidently, Cleveland did look into the matter, because on July 3, Morton announced that Harrington had been exonerated of all charges.

Less publicly now, the frustrated secretary of agriculture and his ally Dunwoody continued their campaign against Harrington and the Weather Bureau scientists. Morton fired two leading researchers, citing the need to cut costs, and reduced Cleveland Abbe's salary. A consoling letter to Abbe came from Gardiner Hubbard, an influential lawyer who with his son-in-law, Alexander Graham Bell, had created the Bell Telephone Company and had purchased and reorganized the journal *Science*. "Morton does not understand the value of scientific investigation of any project," Hubbard wrote. "I shall try to do something for science, which includes you, and find time to talk to the president." In a letter to the editor of the *New York Tribune,* Abbe wrote, "Ever since General Hazen encouraged development of the Study Room, the Army officers foresaw a dangerous rival." Morton continued to praise Dunwoody, who was allowed to make sweeping policy changes while Harrington was away.

Harrington addressed the problem in his 1895 journal article, referring to "some prevalent associations" with the idea of science. "To some 'science' is a monster of frightful mien, bristling with differential equations, parallelograms, paradoxes and other dreadful sesquipedalian things," he wrote. "To others it is like religion, unpractical, and the scientific man . . . is assumed to be an incompetent incapable of running business affairs. Others think that scientific work is not a function of government, their logic apparently being as follows: Some science is without practical application, therefore no science should be fostered by government." As the months of conniving dragged on, the scientist could see his control of the agency and his vision of its mission slipping farther and farther away. In a letter to Morton on April 30, 1895, Harrington finally blurted it out: "Dunwoody is a selfish intriguer and a source of discord in the Weather Bureau. I request that the President recall him. We do not need military control of the Weather Bureau."

This letter apparently proved to be the breaking point in the impasse. On June 19, President Cleveland finally acceded to Morton's appeals and sent word to Harrington that his resignation would be acceptable, citing as grounds "personal interests." Harrington refused to tender it, citing "public interests," and so, effective July 1, he was fired by the president. "Cleveland Wields His Axe, and Professor Harrington's Official Head Falls," headlined the *New York Tribune*. "Partisan politics and science come into collision in the Weather Bureau with disaster to the latter—Secretary Morton obtains his enemy's scalp at last." In a statement, Harrington said,

"Among the public interests which I have had steadily in view were the preservation of the scientific corps and the protection of the bureau from the spoilsman. When a scientific bureau descends to the four-year office-holding plane it at once loses prestige, and ceases to be a desirable post for competent men."

The action was roundly condemned by newspapers around the country. Alluding to Dunwoody's and Morton's attitudes toward research, the *Philadelphia Inquirer* suggested that instead of "having to pass an examination in the science of meteorology, candidates for office in the Weather Bureau will be asked to show their knowledge of the habits of cats, rats, dogs, sheep, wild geese, chickens, cows and horses when the weather is about to change." Cleveland's apologists offered the excuse that the head of the Weather Bureau "ought to be in personal as well as official accord with the head of the Department," the *Tribune* reported. For his part, Morton was disingenuous to the end. "This is a matter that belongs properly to the White House," he said. "I have nothing whatever to say upon the subject."

As the replacement for Harrington, Morton undoubtedly preferred his favorite army officer over a scientist, but Dunwoody declined to abandon his military commission in order to make himself eligible. Cleveland Abbe was the obvious choice within the bureau, but Morton was not about to consider someone who had spoken out against the damage he had done to the bureau's scientific reputation. If Morton had even bothered to look, he would not have found leading scientists outside of government interested in working for him. He dug deep into the ranks of the Weather Bureau and found Willis Luther Moore, who at 39 was the youngest professor in the agency. Moore was not a scientist of Harrington's caliber, but he would prove to be an infinitely better politician.

Within three weeks of his firing, Harrington had before him an offer by the regents of the University of Washington to become president of the institution. He had come highly recommended by President James B. Angell of the University of Michigan as well as David Starr Jordan, president of Stanford University in California, who suggested that perhaps Harrington "has had enough of political forecasting." In August he accepted the offer, and in September he was a continent away from J. Sterling Morton's intrigues, helping to develop an institution of academic distinction in the Pacific Northwest. But the political winds hounded him still. Agrarian populists swept into the state legislature from eastern Washington and undercut the university's political base. While many thought that Harrington's experience as a federal administrator would have prepared him for such a battle, just the reverse seems to have been the case. "Harrington found it difficult to come to grips with university problems, or to decide upon and effect necessary actions," wrote a university historian. Looking on from California, Jordan wondered if Harrington's experience with the Weather Bureau had not ruined him. In March 1897, he resigned.

Now something was wrong. In a letter to a University of Washington friend in December, he asked for news of colleagues and alluded to the difficult and deeply personal nature of the problem: "As to myself—but why expand on that theme—I am still trying to solve the unsolvable." Harrington wrote a book, *About the Weather*, which appeared in 1899. In September 1898, he became director of the Weather Bureau office in San Juan, Puerto Rico, and wrote a magazine article about the island. But his days in science, and in administration, were over, and he was on the verge of complete mental collapse. He was removed from his position for failing to perform his managerial duties. In March 1899, he was transferred to the bureau's New York City office, but failing mental and physical health forced his retirement in June.

Later that year he walked out of his house to attend a dinner and never returned. It was the beginning of a long and solitary odyssey. Its barest details are sketched from what one biographer calls "one or two strangely worded letters and an occasional news item indicating that a learned and cheerful philosopher was working in a lumber camp, on a sugar plantation, or in a shipyard." For nearly 10 years Mark W. Harrington wandered like a ghost over the trail of his distinguished past. Back to Louisiana, now as a plantation worker. Back to China, now in a shipyard. Back to the Pacific Northwest, now in a lumber camp. In June 1907, he showed up at the Second Precinct police station in Newark, New Jersey, asking for shelter. A police officer asked him his name. He said he didn't know it. He was placed in the Morris Plains Asylum for the Insane, where his wife, Rose Smith Harrington, and his son found him in November 1908, number eight among the John Does.

In December 1908, the journal *Science* reprinted this letter to the *Boston Transcript* from a George N. Lovejoy:

> It is extremely difficult to realize the sad termination, in all probability, of the career of Professor Mark W. Harrington, formerly at the head of the astronomical department in Michigan University, latterly chief of the United States Weather Bureau, Washington. One of the brightest intellects and most successful instructors, whose work as a teacher, not only in this country, but in China, years ago, brought him into prominence among scholars everywhere; whose career, though brief, at Washington

> was such as to redound to his credit and the honor of the government; a man of rare conversational gifts, an interesting personality, genial at all times, it is hard, indeed, to realize that such an one to-day . . . has been an inmate (until recently his identity unknown) of an insane asylum, his mind a melancholy blank.

Following this letter in *Science* was an item reporting that a "Cleveland Memorial Association has been formed, its object being to erect in Princeton a suitable memorial of the late President Cleveland."

Harrington's physical and mental state seemed to improve from time to time, and there were times when he had hopes of returning to the scientific and intellectual world. But he never recovered enough to leave the asylum. He died there in 1926.

15

Isaac Monroe Cline

Taking Galveston by Storm

•

ISAAC CLINE KNEW it all. Until September 8, 1900, he seemed to think that everything there was to know about the weather, everything important, already was known, and that he knew it. Questions were few, in Cline's mind; answers were many. Nothing in his utterances or writings, nothing in his upright bearing, gave rise to the slightest doubts about his command of meteorology or the science's understanding of the ways of the atmosphere. An 18-year veteran of the service, Cline had come up through the ranks of the old U.S. Army Signal Corps and had shown himself to be just the kind of man that J. Sterling Morton and Colonel H. H. C. Dunwoody had in mind for the new U.S. Weather Bureau when they finally rid the agency of its first civilian director, the scientist Mark W. Harrington. If Willis L. Moore hadn't wanted the job, Cline would have been an excellent choice. Both men were attuned to the rising commercial strength and the political power of the period. Congress and the country at the turn of the century didn't want to hear from a bunch of equivocating intellectuals about how complicated things were, about what they didn't know and couldn't do. The North American continent had been conquered, Cuba and the Philippines had just been taken from Spain, and in 1900 the United States was ready to take on the world. From its national weather agency it wanted service and certainty. Weather forecasting and the U.S. Weather Bureau were not about scientific research. That mistake would not be made again—not by a politician, not by a leading scientist, and certainly not by Willis Moore. The new Weather Bureau chief found in the local forecast official of the

Galveston, Texas, office a good friend and one of his most trusted lieutenants. Isaac Cline was a man of the hour: industrious, ambitious, assertive, pious, and proud. Most of all, until September 8, 1900, he was sure of himself.

But even in those heady days of Galveston's rapid growth, at least some people once in a while wondered about the city's vulnerability to tropical storms from the Gulf of Mexico. After all, the city was building itself on a long, low sandbar between the gulf and Galveston Bay, a barrier island with its highest point less than nine feet above sea level. To most people, most of the time, this exposure to the sea was one of Galveston's great charms, a prospect that often saved its residents from the worst effects of summer's suffocating humidity. This busy port city was in love with its waterline. Even the natural dunes along the seafront that might provide some modest protection from high tide and surf had been removed to improve the sea views and access to the beach. At the end of long piers, elaborate bathhouses hung out over the gulf. Passengers on arriving steamboats saw a city that seemed to rise from the water like a shimmering mirage. Still, even as the city competed for port business with nearby Houston, there were those who realized that Galveston owed much of its recent growth and prosperity to the catastrophic demise of the once-booming port city of Indianola to the southwest. A hurricane had devastated Indianola in 1875, and another in 1886 had completely wiped it off the map. Since 1886, the question had haunted the city: If Indianola, why not Galveston? In the aftermath of Indianola's demise, a group of Galveston businessmen considered building a protective seawall, but their resolve subsided along with the floodwaters of 1886. In July 1891, another drenching tropical cyclone came along that caused local flooding in Galveston, giving new life to the nagging doubt. The occasion prompted the *Galveston Daily News* to ask the head of the local U.S. Weather Bureau office to evaluate the city's vulnerability.

Cautious meteorologists might have shied away from such an invitation, explaining the difficulty of predicting the paths of future hurricanes. Many would have couched their estimations in the language of uncertainty. But there was nothing shy about 29-year-old Isaac Cline in 1891, and nothing uncertain. He seemed to revel in this opportunity to strut his scientific stuff. Here was Cline at his best, waxing disputatiously on a subject that was then—and remains a century later—one of the most enduring puzzles in meteorology. Why, exactly, do hurricanes go where they go? Everyone in the nineteenth century underestimated the difficulty of answering such a question, just as they underestimated the problems associated with accurate weather forecasts. None were farther from the mark than the men of the old Signal Corps. In tone and content, Cline's confident assessment in 1891 would do more than reassure Galveston's worried souls. It would argue persuasively against the necessity of a seawall at a crucial time in the city's history. It would handsomely boost the expansive appetites of the city's commercial leaders. And certainly it would please U.S. Weather Bureau chief Willis Moore and his superiors in Washington.

Cline's article in the *Daily News* was illustrated with a map of tracks for several tropical cyclones that had moved through the Gulf of Mexico over the preceding 20 years. He observed:

> [O]ut of about twenty West India hurricanes which have at times passed over the southern coast of the United States, only two have reached Texas, and out of a large number of cyclonic disturbances which have had their origin in the gulf of Mexico and seven particularly in the west gulf, only two have passed over the Texas coast. The coast of Texas is according to the general laws of the motion of the atmosphere exempt from West India hurricanes and the two which have reached it followed an abnormal path which can only be attributed to causes known in meteorology as accidental.

Likewise, the storms of the west gulf "are generally developed about the parallel where tropical storms change their courses to the north and northeast, consequently their general course is northeasterly and while parallel to the coast of the state are so far away that damaging effects are seldom experienced."

Where in the world, where in the literature, did the scientist in charge of the Galveston Weather Bureau office find "causes known in meteorology as accidental"? Was it something that Matthew Fontaine Maury had written? Coincidentally, or not so coincidentally, it was Maury, writing in *The Physical Geography of the Sea,* who was equally confidant that a place like Galveston would be safe from inundation during a tropical storm because the long, shallow beach in front of the city would cause the storm-charged surf to crash and spend itself before reaching the shore. Cline also argued that other lower-lying land on the nearby mainland would absorb the floodwaters before Galveston was seriously affected. Cline delcared:

> The Texas coast . . . is much less subject to severe meteorological disturbances than many other portions of the country. The opinion held by some who are unacquainted with the actual conditions of things, that

> Galveston will at some time be seriously damaged by some such disturbance, is simply an absurd delusion and can only have its origin in the imagination and not from reasoning; as there is too large a territory to the north which is lower than the island, over which the water may spread, it would be impossible for any cyclone to create a storm wave which could materially injure the city.

Having liberated the unreasonable and uninformed worriers of their absurd delusion, Cline went on to say that because of the trajectory of storms and the lay of the land, the city did not even have to worry about accidents of nature. "Galveston, on account of its location and the peculiar features of the gulf coast, is not liable to be caught in the track of any West India hurricane which might from accidental causes be carried into the west gulf."

Nothing Isaac Cline wrote in 1891 would in any way explain what happened on Saturday, September 8, 1900. In every important respect, Cline and everyone at the U.S. Weather Bureau was generally wrong about the behavior of hurricanes, fundamentally wrong about the vulnerability of Galveston, and specifically wrong about the location and character of a particular tropical cyclone that traveled from the Caribbean Sea through the Gulf of Mexico the first week of September.

This last disastrous failure was in significant measure the result of an act of singular political arrogance on the part of Moore and Colonel Dunwoody. Just days before the Galveston hurricane, the chief of the U.S. Weather Bureau, at Dunwoody's urging, persuaded the War Department to deny access to Cuba's own telegraph system the weather warning cables that had long been issued by the Meteorological Observatory of the Royal College of Belen in Havana. Under the late Jesuit priest Father Benito Vines, the Belen observatory had been studying hurricanes for 30 years and was an eminent meteorological institution. Greatly annoyed by the independent weather forecasts from Belen, Moore and Dunwoody argued that its hurricane warnings unnecessarily riled "the natives." The Cubans were enraged by this action and the telegraph ban eventually was rescinded, but not before Moore and Dunwoody had successfully blocked from telegraphic distribution the only warning in the first week of September 1900 that a tropical cyclone was heading into the Gulf of Mexico and was aimed for Texas.

Nothing that Moore and Dunwoody expected of the hurricane came to pass. While it was barreling through the gulf, Moore placed it near Florida and warned that it would travel up the Atlantic coast. After it made landfall in Texas, and before it ripped up through the Midwest and arced far north out across southeastern Canada for the better part of a week, Moore predicted that the hurricane would dissipate over Oklahoma and no longer pose a threat.

From the time it entered the Gulf of Mexico until the city was flooded by its storm surge and shredded by its winds on the night of September 8, neither Willis Moore from Washington nor Isaac Cline on the scene warned anyone in Galveston that a hurricane was on the way. That their blunders did not provoke a national scandal, that their heads did not roll, can be attributed to the fact that the perpetrators were in the fortuitous position of telling the rest of the country the story of what happened that day in Galveston—and to the fact that Isaac Cline's experience was so harrowing, and his personal loss so great.

"By 8 P.M. a number of houses had drifted up and lodged to the east and southeast of my residence, and these with the force of the waves acted as a battering ram against which it was impossible for any building to stand for any length of time, and at 8:30 P.M. my residence went down with about fifty persons who had sought it for safety, and all but eighteen were hurled into eternity," he wrote in a special report to headquarters on September 23. "Among the lost was my wife, who never rose above the water after the wreck of the building." Clinging to floating debris, Cline managed to save himself and his two young daughters. Several days later, the body of Cora May Cline, pregnant with a third child, "a beautiful, brilliant and cultured girl," the love of his life, was found tangled in the debris of the house.

What agony must have visited a man so singularly responsible for miscalculating the danger to his family and friends and the city of Galveston. Publicly at least, Isaac Cline never let on. Whatever haunting doubts he might have had about his role in the most deadly weather disaster to befall his country, Isaac Cline took to his grave in New Orleans in 1955 at the age of 93. Meanwhile, by September 23, 1900, Cline was back on his feet sufficiently to carefully misrepresent what had happened.

"Storm warnings were timely and received a wide distribution not only in Galveston but throughout the coast region," Cline wrote in his special report to his boss, Willis Moore. Years later, in his autobiography, Cline would state that "neither hurricane nor emergency warnings were received" from Washington before the storm struck and would describe how he had so cleverly covered up the fact. He wrote:

> Some days after the disaster the local representative of the Associated Press came to me with a telegram from his President in New York asking him if hurricane warnings had been displayed. I told him that storm warnings were displayed for two days previous to the hurricane, and that he knew of the urgent warnings and advices I had given the people

> and how I had gone among them early on the morning of the 8th and told them of their danger. I then asked him if he thought any thing more could have been done to warn the people, and he replied, "Nothing more could have been done than was done." I then said, "Telegraph that to your President," which he did and that closed the inquiry without involving the forecaster at the Central Office from whom no hurricane warnings were received.

Everything Isaac Cline reported was calculated to safeguard the reputations of the Weather Bureau, Willis Moore, and, of course, himself. Out of his own account, Cline emerged a hero. "The public was warned, over the telephone and verbally, that the . . . worst was yet to come," he wrote. "People were advised to seek secure places for the night. As a result thousands of people who lived near the beach or in small houses moved their families into the center of the city and were thus saved." Later versions of the story would have him riding up and down the beach that morning, hailing everyone within reach of his voice, warning vacationers to go home. In the years following the disaster, in all of the memoirs of the thousands of survivors, no one besides Isaac Cline seems ever to have made note of hearing or heeding those warnings that morning, or of witnessing that heroic act. By some arithmetic that he never explained, Cline would estimate that he had saved at least 6,000 lives. Was it coincidence that at the time the toll of the storm was estimated at 6,000 dead? The actual death toll was something that neither Cline nor anyone else would ever know. Whole families disappeared, leaving no one to look for them. Rotting bodies were piled and burned, and untold hundreds or thousands drifted out to sea. In a widely publicized letter in response to Cline's report, Moore said, "The record shows that you were all alert and vigilant from the time the first notice of the storm was received," that thousands of people had been saved by his warnings, and that he showed "courage and fidelity to duty that every employee of the Bureau should be proud to emulate."

Whatever its personal impact, the Galveston hurricane changed Isaac Cline's professional goals. He had earned a medical degree in his "recreation time," as he called it, and he had been studying the relation of climate and weather to human health when the storm struck. In fact, a book manuscript on the subject was lost in the wreckage of his house. Cline abandoned that effort in favor of research into tropical cyclones. He wrote years later in his autobiography: "The destruction wrought by the storm tide caused by the winds of the cyclone, and the appalling loss of life in other tropical cyclones from the same cause convinced me that with proper knowledge as to the cause of these storm tides the tremendous loss of life and property could in a great measure be prevented." He devoted many years of his life to this line of investigation and became an internationally recognized research meteorologist. He became a more thoughtful and more serious scientist, perhaps even less insufferably pompous in his pose, although Cline would never be known for humility.

Along the way, he would get a taste of his own medicine. Among the scientists at the Weather Bureau, he encountered a wall of arrogant opposition to his new theories on the structure of hurricanes, the configuration of their winds, and the storm surges they generate. Cline's work challenged, and eventually overturned, prevailing descriptions of tropical cyclones. Cline divided the storms into quadrants—front and rear, left and right—and showed how different forces alter the course and power of winds and waves around the storm. Where conventional descriptions showed winds rotating uniformly around a storm's center, Cline showed a different pattern in each quadrant and how the strongest winds are in the right rear, where the winds are moving in the same direction as the storm itself. Where conventional descriptions illustrated storm tides radiating evenly outward from its center, Cline showed that the surge is concentrated in the right front quadrant, propelled far ahead of the storm by the strongest winds behind it. His work on the subject eventually was published over the objections of the scientists entrenched in the Weather Bureau, although he observed that none of them ever published a formal scientific challenge to his research.

In 1926, he published *Tropical Cyclones,* a book that became an important text on the subject. In 1934, Cline served as president of the American Meteorological Society. In the book, and in a presidential address to the organization, Cline paid tribute to the fact that William Ferrel, 50 years earlier, had correctly described the nonuniform structure of winds in cyclones, but that nobody in 1885 had taken notice. Cline had come a long way from the days when people who held views different from his own were suffering from "absurd delusions." "The reason Ferrel's great contribution to the study of cyclones was not accepted can only be attributed to the petty jealousy of fellow scientists and investigators in the study of meteorology," he wrote.

Willis Moore's political appetite eventually caught up with him. In 1912, he became the subject of an investigation into the misuse of government resources in his unsuccessful campaign to be named secretary of agriculture. He was fired.

For many years now, when employees of the National Weather Service are recognized for exemplary service to the agency, they are given the Isaac Cline Award.

16

Gilbert Walker

The Southern Oscillation

•

WE LIVE IN A WORLD of stirring fluids—water and air, ocean and atmosphere. This is what modern climate science has discovered: that the thick, slow water and the thin, fast air are intimately bound to one another, always mixing and exchanging energy, like two very different individuals carrying on some interminable conversation. To one particularly emphatic point in the conversation, when sea and sky seem to be shouting at one another, we now give a special name: El Niño. Often as not, among themselves scientists will use the term *El Niño/Southern Oscillation* (ENSO) to more faithfully describe both sides of this exchange. As it happened, the atmosphere's voice, the Southern Oscillation, was the first side of the conversation to be heard by modern science.

It happened in turn-of-the-century India—imperial British India—in the town of Simla, the colonial summer capital nestled in the green foothills of the Himalayas. Simla was what was known as a "hill station," being located up above the subtropical inconveniences of the subcontinent. Often a solitary figure could be seen out on the flat, manicured lawn at Annandale, the parade grounds where Britannia strutted its stuff for the viceroy and his military friends. Except for the object in his hand, a boomerang, he was the perfect image of the starched Cambridge University don. He was lean and erect, his bearing that of a man accustomed to quiet authority. It was 1904, and at the age of 36, the man on the big lawn, Gilbert Thomas Walker, had taken the post of director-general of Indian observatories—India's chief meteorologist. Only recently had he arrived from Cambridge, from the cloistered security of a lectureship at

Trinity College. He was a scientist, a scholar, and here he looked every bit the part. Well, except for the boomerang.

In an age of classical conformity, in a regimental outpost of empire, here was a very original man. First and foremost, he was a brilliant mathematician—"a mathematician to his fingertips," in the words of a peer. But Gilbert Walker was also a painter of watercolors and a musician—an accomplished flutist. He was an expert ice skater, an indefatigable mountaineer, a naturalist, and a birder. He was an astute conversationalist, a fine friend, by all accounts, and a man of uncommonly good disposition. Many years later the *Times* of London would write how the "breadth of his interests, the lucidity of his conversation and the sweet reasonableness of his nature made him an agreeable figure in Simla society and in London scientific circles." And without a doubt, in his day he was the best boomerang thrower in all of Europe and India. So there he would be, out on the lawn of the parade grounds, which were about the size of a football field and the only level ground in the Simla hills, and so the only really suitable place for throwing boomerangs.

Gilbert Walker was first to define an *oscillation,* a fluctuating pattern, between the air pressure on one side of the Pacific Ocean and the air pressure on the other. When the air pressure is up on one side, it is down on the other, like a seesaw. While reporting this big swaying of the atmosphere, as he called it, Walker pointed out that the Southern Oscillation seemed to be linked to differences in seasonal temperature and rainfall patterns in far-flung regions of the world. This novel idea he developed in a remarkable series of technical papers titled "World Weather." Published throughout the 1920s and 1930s, these papers first observed the scope and character of year-to-year global climate differences associated with the Southern Oscillation: changes in the winds and rainfall patterns across the tropical Pacific and Indian Oceans; and fluctuations in temperatures over western Canada, the southeastern United States, and southeastern Africa. With this data, Walker pointed to the possibility of long-range forecasting, or "seasonal foreshadowing" as he called it. The science of climate prediction, known as *seasonal forecasting,* begins with the Southern Oscillation and these papers. The science of the day was not ready to take on the problem of world weather, however, and recognition of his accomplishment would be a long time coming.

Gilbert Walker was an altogether improbable candidate to head up India's sprawling meteorological service in 1904. He was recognized in London and Cambridge scholarly circles as a mathematician of singular ability and had achieved high academic honors. But his field was mathematical physics, and his chief work—in addition to the dynamics of the flight of the boomerang, of course—was in the mathematical theory of electricity and magnetism. His paper "Repulsion and Rotation Produced by Alternating Electric Currents" was read before the Royal Society in 1891, and his 1897 treatise on the boomerang was widely regarded as a highly original and mathematically innovative work. But Walker had not read a word of meteorology when Sir John Eliot returned to England in 1903 in search of a successor as chief meteorological reporter to the government of India and director-general of Indian observatories. Moreover, he had no administrative experience or training at Trinity College.

The situation in India at the time was desperate and beyond the experience of European meteorology. Millions had perished in the famines of 1877 and 1899, in droughts that followed the failure of the monsoons, and the subcontinent urgently sought an understanding of these seasonal rains on which its agriculture and its people lived and died. Faced with such a demand, Eliot had taken monsoon forecasting far out on a limb. "It is regrettable that under his direction the monsoon forecasts became more and more ambitious, though based on a slender foundation," wrote another British meteorologist, Sir Charles Normand, years later. The forecasts ran on for 30 pages, in great detail, and were often completely wrong. As Normand observed, "Newspapers became rather scathing in their comments."

This was not a position that an average scientist with a reputation to protect would find attractive. The scientific merit of any kind of weather forecasting still was the subject of debate in Europe at the turn of the century, and the idea of long-range forecasting of monsoons, a world away, must have seemed beyond the pale. The British weather service, such as it was, had only recently resumed the practice of issuing forecasts after an abortive beginning under Robert FitzRoy in the 1860s. As criticism mounted in India, the government allowed Eliot to issue his last monsoon forecasts as "confidential" government documents. And when he argued that his successor would need more scientific staff, the officials of the colonial government were inclined to agree with him.

At Cambridge, the new head of India's forecasting services had been christened "Boomerang Walker," because he could so often be seen tossing those ancient aboriginal implements out in the "Cambridge Backs." If Walker had a complaint about life at Trinity College, it probably was its lack of consequential challenge. A contemporary, Geoffrey I. Taylor, observed that in conversation "Walker often commented on the difficulty that applied mathematicians experienced in those days in Cambridge in finding something to work on, owing to the fact that their training did not involve any laboratory work. His attitude to mathematics is well expressed in a remark which he quoted as having been made to him by

John Hopkinson: 'Mathematics is a very good tool but a very bad master.'" What Eliot, a fellow Cambridge scholar, offered Gilbert Walker was something to work on.

"It was a surprising, but a very wise choice," wrote a colleague. "He chose a man full of energy, ready to learn and who proved to be a man capable of making personal friends of his colleagues, both British and Indian."

Meteorology in India, especially monsoon forecasting, needed building from the ground up. In the late 1880s, the Indian Famine Commission had asked Henry F. Blanford, the colony's first meteorological reporter, if science could foresee variations in the monsoon rains that irrigate India from June through September. Blanford thought the rains were driven by relatively local conditions, such as the amount of snow in the Himalayas, the summer heat or winter cold in central Asia and Tibet, and storms in Persia. His forecasts were not very successful, however, and under Sir John Eliot things had gone from bad to worse. Eliot seems to have tried to make up for what his science lacked by providing voluminous detail delivered in what Walker described as "a somewhat pontifical" mode of expression.

Boomerang Walker did not bring a pontifical mode of expression with him to India. He brought a boomerang, a fresh mind, and a completely new approach. If Blanford's regional factors weren't responsible for monsoon variations, the answer must lie farther away. New to the subject, Walker was free to consider possibilities that scientists steeped in the conventional meteorological thinking of the times might have rejected. He would not look for an explanation of the monsoons, but rather a device to render a forecast. He would look for a sign elsewhere in the world that was linked to their occurrence, for whatever reason. He would cast his net as far as possible, gathering weather records wherever in the world he could find them. Walker obtained pressure, temperature, and rainfall records, usually of around 40 years duration. He went beyond the records of traditional meteorological parameters, including time series of such miscellaneous variables as river flood stages, mountain snowpack depths, lake levels, and sunspot activity in an effort to detect cause-and-effect relationships.

Unique to the science, Walker's method of research would apply his strengths as a mathematician. The meteorological would become numerical; his search for clues to the monsoon would be constituted as a monumental problem of statistics. Each weather record was averaged by season within each year and then grouped by season to form four time series for each variable at each station. According to an obituary in the

Times of London, "an early visitor to his office at Simla has recorded a vision of 'rows and rows of pigeonholes' in which 'all sorts of curious coincidences' were recorded. When Walker found two phenomena, varying in unison over a series of years he put them into one pigeonhole, and eventually into one diagram, though there was no apparent causal connection between them."

He employed a method of statistical analysis that produces what is known as a *coefficient of correlation,* a single numerical measure of the closeness of the relationship between any two factors. It is rigorously objective. Two factors in perfect agreement with one another achieve a value of 1.0, while complete disagreement produces zero. When two are in perfect agreement but of opposite phase—a seesaw pattern—their relation is expressed as a value of –1.0. Walker and a raft of assistants compiled voluminous tables of contemporaneous, serial, and time-lag correlation coefficients. It was difficult and tedious work, involving hundreds of hours of calculations by scores of assistants. In the hands of a lesser mathematician, it could have been folly; but Walker knew what he was doing. Where a method did not exist to handle a particular statistical problem, Walker would invent one.

"Walker's work was very innovative on the statistics side," noted Richard W. Katz, a mathematician and senior scientist at the National Center for Atmospheric Research in Boulder, Colorado. "Methods like time-series analysis, even certain sophisticated aspects of regression correlation, were just getting going when Walker was starting to do this work." To statisticians, the name Walker is applied to the so-called Yule-Walker recursion, a set of equations used to discern periods in time-series analysis. To meteorologists, Walker is known for the Southern Oscillation. "Nobody seems to realize it's the same person," said Katz. More remarkable, perhaps, the distinctions in both disciplines are the result of the same research.

"The main conclusion," Walker reported in 1928 to the Royal Meteorological Society, "is that there are three big swayings or surgings: (a) The North Atlantic oscillation of pressure between the Azores or Vienna on one hand and Iceland or Greenland on the other; (b) The North Pacific oscillation between the high pressure belt and the winter depression near the Aleutian Islands; and (c) The southern oscillation, mainly between the South Pacific and the land areas round the Indian Ocean." It was an unfortunate choice of terms, in a way. While the North Atlantic and the North Pacific accurately imply regional effects, the term *Southern Oscillation* completely misses the more powerful character of this climate feature and the global extent of its reach.

Walker himself had no illusions about this. "The southern oscillation is more far-reaching than the two oscillations just described," he said, "and as the effect of an abnormal season is propagated slowly, it may not appear at the other side of the earth until after an interval of six months or more." Here was the key to forecasting the monsoon, he said, as well as other far-flung features of seasonal weather. "High pressure in the Pacific and South America is associated with low pressure in the lands round the Indian Ocean, with low temperatures in tropical regions and the centre of North America, and with abundant rain in India, Java, and Australia, and high Nile floods, while the rainfall is scanty in Chile." Rainfall in Hawaii was part of the phenomenon, he said, as were winter temperatures in western Canada.

Walker had what he was looking for: an objective basis for a formula to forecast Indian monsoon rainfall. "Examination by statistical methods," he said, "has brought to light many relationships between seasonal conditions in different parts of the world, usually contemporary but often three months or more apart; and in the latter case knowledge of the earlier conditions in one region may give a rough idea of what will occur later on in the other."

His research might have served as an outline to lead the young science of meteorology toward the grasp of climate and weather as global-scale events. Certainly it represented a new direction for the science. Instead, Walker's methods drew criticism from leading meteorologists in London. This Cambridge mathematician may have thought he was out doing operational fieldwork, but it bore little resemblance to the treatment conventional researchers were accustomed to seeing. His statistical analysis seemed entirely blind to dynamical theory and was outside of science's time-honored standard model: form a hypothesis, an idea, and then go test it.

Walker was compiling statistics from naked data and drawing ambitious conclusions about the possibility of long-range weather forecasting on the basis of far-flung numerical relationships. He couldn't explain these relationships, and neither could the theorists. Walker's approach was described as entirely *empirical* at a time when the term was, as one writer put it, "a heavy missile in the mouth of a scientist."

"There is no *a priori* reason to expect this relationship between two events so far apart in space and time," noted Sir Charles Normand. Walker sought a physical explanation for these global weather relationships by relating them to sunspots and to sea ice in the South Atlantic, but the results were not conclusive; the correlation coefficients were not high. Except in cases of extremely high correlations, said the British meteorologist William H. Dines, "forecasts based on correlation coefficients . . . can hardly be much more than pure guesses." In London, the conventional course would have been to use correlations to guide the search for a theory upon which to base forecasts. In India, Gilbert Walker, who was not a theorist in any case, did not feel that he had the luxury of waiting. As Normand later observed, "To act upon the principle of a theoretical basis would have postponed all further attempts to advance seasonal prediction.

"In the circumstances, Walker frankly followed empirical methods, arguing that if, as seems to be the case, physically real relationships can be found by strict statistical methods, ignorance of their explanation should not stop him from using them, and that the more relationships we find empirically the more insight we shall have into the physical facts of the problem, and the more likely are we to find some ground on which to base a theory."

Ironically, the relationship of the Southern Oscillation to India's southwest monsoon proved to be weaker than to most other seasonal conditions Walker identified. Over time, his monsoon forecasting formula has become less reliable, in part because India's climate has changed. A century later, it still puzzles climate scientists.

Speaking to the Royal Meteorological Society in 1953, Normand observed that the monsoon rainfall "plays a part" in Walker's Southern Oscillation:

> Unfortunately for India, the southern oscillation in June-August—at the height of the monsoon—has many significant correlations with later events and relatively few with earlier events. To my mind, the most remarkable of Walker's results was his discovery of the control that the Southern Oscillation seemingly exerted upon subsequent events and in particular of the fact that the index for the Southern Oscillation as a whole for the summer quarter June-August, had a correlation coefficient of +0.8 with the same index for the following winter quarter, though only of –0.2 with the previous winter quarter. It is quite in keeping with this that the Indian monsoon rainfall has its connections with later rather than with earlier events. The Indian monsoon therefore stands out as an active, not a passive feature in world weather, more efficient as a broadcasting tool than as an event to be forecast.

Normand noted the irony: "On the whole Walker's world-wide survey ended by offering more promise for the prediction of events in other regions than in India."

But Walker was one of only a few researchers thinking in terms of global forecasting at the time. And he lacked critical data. What was going on in the upper reaches of the atmosphere above the seesawing surface pressures he discerned? What was going on in the ocean? There was virtually no data on the upper air or on sea-surface temperatures of the vast expanse of vacant Pacific Ocean between the hemispheres. Without this basic information he could not make a coherent theory of world weather, even though he could put his finger on its rhythms. In 1918, he wrote: "I cannot help believe that we shall gradually find out the physical mechanism by which these are maintained, as well as learn to make long-range forecasts to an increasing extent."

Despite a distinguished career, Walker would not see the day when his most important contribution to climate science would be recognized. For his service in India, he would be knighted on his retirement in 1924. Walker would become Sir Gilbert. He would move on to a prestigious professorship in meteorology at the Imperial College of Science and Technology in London. He would be made a fellow of London's learned Royal Society and serve as president of the Royal Meteorological Society. Other scholarly accolades would come his way. But his greatest scientific achievement would go largely unheralded in his lifetime. Walker would live until November 4, 1958, four months past the age of 90, and still it would not be long enough. Writing his obituary in 1959, a fellow meteorologist remarked on the long series of "World Weather" papers: "Walker's hope was presumably not only to unearth relations useful for forecasting but to discover sufficient and sufficiently important relations to provide a productive starting point for a theory of world weather. It hardly appears to be working out like that."

Climate science would go off in other important directions over the years. If the Southern Oscillation was observed at all, it was seen as a curious, statistically derived collection of coincidences that nobody could explain: rainfall up here; pressure down there; temperatures rising and falling according to some unknown rhythm, or maybe nothing at all. Decades were to elapse before a brilliant theorist would come along and put it together. In the 1960s, the Norwegian Jacob Bjerknes would connect what Gilbert Walker found in the atmosphere to what others were finding in the ocean, and finally give meaning to the global reach of the El Niño/Southern Oscillation.

Eventually, it all would come back to the Southern Oscillation and to the statistical methods pioneered by Sir Gilbert Walker in Simla, like a boomerang after a long and particularly interesting flight. The day would come when scientists in a powerful new field would realize their debt to him. The Southern Oscillation Index would become a standard tool in modern climate science and a key to seasonal forecasting. The relationships Walker defined, which seemed so inexplicable in the early days of the twentieth century, would stand the test of time, and modern scientists would make a vigorous new discipline of finding such *teleconnections* between the Southern Oscillation and world weather events.

In Sir Gilbert's day, El Niño was a name Peruvian fishermen gave to a relatively small circumstance, the occasional appearance of a seasonal countercurrent that appeared off their shores. The anchovy would leave, and they would take a few months off. Years later, when the phenomenon took on its larger meaning, its greater extent, El Niño would be understood in the context of the Southern Oscillation. Near the turn of another century, when Ants Leetmaa, then director of the U.S. Climate Prediction Center, defined the closeness of the relationship between El Niño and the Southern Oscillation, nobody would have more clearly understood or deeply appreciated the terms he chose to make the point. "They correlate at about .995," Leetmaa would say, "which means they are virtually the same signal."

17

C. LeRoy Meisinger

Death by Daring

•

"THERE IS A VERY decided charm about being suspended in the air, especially when one is uncertain as to his location, and when one does not know his speed," Lieutenant C. LeRoy Meisinger of the U.S. Army Signal Corps' Meteorological Service reported in the U.S. Weather Bureau's *Monthly Weather Review* a month after ascending in a balloon from Fort Omaha, Nebraska, on March 14, 1919. Meisinger was a talented writer, a fact that found its way even into the stodgy gray pages of a government scientific journal. With the rising sun, "we were permitted to witness such a scene of beauty as no one can dream of who has not explored the wonders of the upper air." Below the balloon was a "gently undulating sea of fog, soft as down and delicately tinted as mother-of-pearl," and overhead floated "a layer of alto-cumuli opaled by the rising sun, and ever varying in iridescent splendor." Enraptured as he was, this uncommonly artistic and creative young man was a dedicated scientist, and his visit to the wonders of the upper air was a singular and timely mission. "This journey did more than provide those thrills; it enabled us to actually penetrate and become a part of the wind circulation of a strong cyclone."

Out of the muddy nightmare of World War I, the old dream of the conquest of the air was becoming a transforming reality. Since the signing of the armistice in January, a vision of civilian aviation was beginning to take hold across Europe and the United States. The benevolent skies of the near future would contain all manner of aircraft: balloons, dirigibles, and airplanes, each used according to their special advantages. A boundless and boundlessly enriching age was dawning. Celebrity daredevils were competing in long-distance balloon races that were major national and international sporting events. Enormous rigid airships and buzzing little airplanes were increasingly common sights over major cities. From the new weapons of the Great War could be fashioned a great new peace: all nations soon would be joined by systems of aerial transportation that would shrink space with their range and time itself with their speed. What remained was to make these vehicles safe and their systems reliable. And investors everywhere were willing to believe that for the same engineering genius that had created these marvelous machines, such improvements were just a matter of time.

Quite apart from the engineering challenges, however, civil aviation envisioned a level of scientific meteorological services that did not exist. Weather science's rudimentary grasp of the atmosphere above the surface of the earth was beginning to be recognized as a serious obstacle to progress. As a licensed balloon pilot and a student of meteorology, Meisinger was acutely aware of the gap between what balloonists and the pilots of dirigibles and airplanes required practically and what weather forecasters of the day could reliably provide them. As a service and a science, meteorology was about to be yanked in a whole new direction. As it happened, few people in the United States, and certainly no one in the U.S. Weather Bureau, was as well positioned, as able, or as eager as 23-year-old Clarence LeRoy Meisinger to try to fill the scientific void created by the emerging development of aeronautics.

LeRoy Meisinger was born in 1895 in Plattsmouth, Nebraska, and he grew up in the state, mostly in Lincoln. He graduated from the University of Nebraska in 1917 with a degree in astronomy. He was a remarkably talented, hardworking, and clean-living son of the Middle West. In addition to the sciences, his university interests ranged through music and art and literature, although not so far into the social realm as fraternity membership, which attracted most Cornhuskers. Nobody said he wasn't friendly, and certainly he was active; but he always seemed more serious than most students around him. He composed music for a drama club and played French horn in the military band. He was a staff artist for the yearbook and a second lieutenant in the university's Reserve Officer Training Corps.

The United States entered World War I in April 1917. Meisinger graduated in May, and in June he joined the army. First he was assigned to the 134th Infantry band, where he played the French horn, but the following April he transferred to the new Meteorological Service of the Signal Corps. He was sent to a special three-month meteorological training course at the Agricultural and Mechanical College of Texas at College Station, where he immediately stood out among the 300 soldiers in his class. "Army associations were responsible for the average man becoming temporarily considerable of a roughneck, but in that particular he was a marked exception to the rule," wrote Ivan Ray Tannehill, a classmate. "Like the majority of men of that type he could do a prodigious amount of work without fuss or confusion." Here the discussions by instructor Charles F. Brooks, a Weather Bureau veteran, of problems relating to the upper air first caught Meisinger's interest. One of only four soldiers to receive commissions after training (along with Tannehill), Meisinger was assigned to inspect the Signal Corps' 37 upper-air monitoring stations in the United States, and then he was appointed chief weather officer at Fort Omaha, Nebraska, where the army housed its Balloon Training School. Here he trained for his license as a balloon pilot and began his investigations of the upper air.

The March 14 ascent from Fort Omaha was typical of scientific balloon expeditions. Free balloon flight and its invitation to the mysteries of the upper air had been beckoning a few courageous European and American scientists for decades. As a platform for science, however, the balloon's inherent lack of control, which young Meisinger found so charming, was not really congenial to many carefully planned experiments of weather research or to the proper care and maintenance of the instruments they required. Drifting and falling through dense fog, blissfully unaware that the altimeter apparently had been damaged during the bumpy takeoff, the balloon carrying Meisinger and crew flew treacherously close to the "squeak-creaking" sound of a windmill that they never saw. In characteristic style, the meteorologist forced himself to confess publicly that through the dense fog he had mistaken the sight of a fast-approaching snowbank for a layer of clouds, "adding with the foresight of his profession that it was unusual to see that type of clouds so low." They nearly crashed, upending a fence post before they threw off ballast and regained altitude. But Meisinger always seemed ready to forgive a free-flying balloon these practical shortcomings. As he wrote in the April 1919 issue of *Monthly Weather Review,* "Owing to a somewhat gusty east wind, which seemed to bear the balloon down upon the ground, the oscillations of the bag were so severe as to cause the basket to crash into the ground on the getaway, rendering the barograph useless. Nevertheless, the whole experience was one of beauty."

A month later, Meisinger again lifted off from Fort Omaha as part of a two-balloon mission suggested by William J. Humphreys, a professor of meteorology at the Weather Bureau, and Willis Ray Gregg, a leading meteorologist at the agency. As Gregg described in a memorandum, one balloon was to maintain a constant altitude at 5,000 feet, the other at 10,000 feet, in order to "enable us to determine the actual trajectory of a particle of air under the influence of a given pressure distribution." Drifting southeastward, Meisinger, in the upper balloon, was able to maintain a constant height, more or less, but the lower balloon got caught up in the turbulent currents over the Ozarks and at one time ended up at 13,000 feet. Still, Meisinger reported that the balloons had been able to maintain their assigned elevations during most of the flight and that the mission was a success. "The paths of both balloons were practically the same, showing that the effect of the controlling cyclone, the center of which was over the Great Lakes, persisted to a considerable height," he stated in August in the *Monthly Weather Review.* "The speed of the upper balloon was slightly greater than that of the lower."

Preparing to leave the Signal Corps in July 1919, six months after the armistice, Meisinger had three interesting career choices. He could do graduate work in astronomy at the University of California; he could become a composer and orchestra leader; or he could join the U.S. Weather Bureau as a research meteorologist and assistant editor of the agency's *Monthly Weather Review,* which in 1919 was being edited by Brooks. However unlikely the choice of the government weather agency may have seemed for an ambitious 24-year-old man of Meisinger's multiple talents, it so happened that he selected the one alternative that held out the possibility of his continuing to fly balloons.

Reporting for duty in Washington in September 1919, this young, bright rising star must have felt profoundly out of place. How Meisinger personally reacted to the stagnating culture of the bureau was something that he had the good sense not to commit to print. Certainly he noticed the institutional isolation of the few scientists in the agency. They were respectfully tolerated, at least formally, but power and responsibility and the road to advancement were elsewhere. No one would confuse the role or the status of the researchers with the people who were doing the actual work of the agency, the weather forecasters, who were not scientists or even would-be scientists.

Young men who hoped to become weather forecasters were expected to have acquired science education only to the level of high school physics. Then they were required to spend several years working at the side of a journeyman, slowly advancing through various levels of junior ranks, gradually absorbing the forecaster's mastery of the weather map and the benefits of his great experience and intuition. "The consensus of opinion seems to be that the only road to successful forecasting lies in the patient and consistent study of the daily weather map," concluded a study of the subject, *Weather Forecasting in the United States,* published by the Bureau in 1916. Professor Humphreys, writing from the safe remove of old age and retirement, would describe this stagnating culture in his autobiography, *Of Me.* According to Humphreys, the bureaucrats running the agency when LeRoy Meisinger reported for duty as a research meteorologist viewed the Weather Bureau wholly as "a service organization to issue forecasts according to empirical rules for this or that type of weather map as evolved by each forecaster for himself. In such an environment collegiate training is apt to be regarded with ill-concealed contempt by those who do not have it, libraries as stores of waste paper, and anyone engaged in real research as at best a worthless idler."

Soon after joining the bureau, Meisinger enrolled in George Washington University's School of Graduate Studies, where Humphreys taught a course in meteorological physics. In 1920, Meisinger received a master of science degree, and in 1922 he earned a doctorate. His thesis, "The Preparation and Significance of Free-Air Pressure Maps for the Central and Eastern United States," was inspired by aviation's need for upper-air wind information, rain or shine. It was a meticulous study of how upper-air pressure patterns could be extrapolated from surface observations by a mathematical process now known as *regression analysis.* The idea was to overcome the fact that upper-air observations by piloted balloons and kites were not available during bad weather—in snow, rain, or fog—when aviators needed them most. "Yet aerial traffic must not be delayed!" Meisinger declared in an article describing the need for the charts. The bureau published the entire work as a special supplement to the *Monthly Weather Review,* and Meisinger was well on his way to being recognized as the country's leading aeronautical meteorologist.

Reviewing the beginnings of civil aviation in the United States in the first year after World War I, Meisinger wrote in 1920 that "the functions of a governmental meteorological agency must be twofold with respect to aeronautics. It must collect and disseminate meteorological data over the entire country, and it must conduct research which will eventually have their reflection in increased accuracy of forecasting." Veterans in the bureau did not share his ambitious vision of meteorology's role in civil aviation any more than they did the young balloonist's enthusiasm for the charms of the upper air. Civil aviation was not *their* dream, and these unrealistic new demands coming from upstart young aviators were not an especially welcome turn of events. The veterans of the bureau had grown accustomed to supplying agricultural and shipping interests with generalized information that everyone knew was not very reliable, but was about as good as it ever was going to get. What aviators wanted were fast, frequent, and reliable forecasts of new weather details such as visibility, upper-air winds, and storminess at specific times and places that were beyond the capacity of their art.

Whatever ambitions the Weather Bureau veterans might have held, they had long since made peace with the limits of their methods. Since the 1880s, forecasters had been tracking changes in air pressure across the landscape; and 40 years later, it was still the basis of their daily predictions. As the 1916 bureau weather study declared, there remained but "two fundamentals of modern weather forecasting. These are (1) weather travels; (2) the character of the weather is in general largely determined by the atmospheric pressure distribution." Nothing more could be gleaned from the all-important *highs* and *lows* on their weather maps. There was nothing more to learn.

It would not have taken LeRoy Meisinger very long to realize that while in the *Monthly Weather Review* he could showcase the important scientific advances of the era—none of that had anything to do with how Weather Bureau forecasters did their work. In February 1919, *Monthly Weather Review* reprinted the seminal papers of Norweigan scientists that defined the critical features of midlatitude storms. The model of air masses doing battle along warm-front and cold-front boundaries was changing meteorological thinking. But not at the U.S. Weather Bureau. In a following issue of the journal, a bureau meteorologist had summarily dismissed the Norwegian model as impractical and not applicable to North American weather. Just the opposite should have been evident, in fact. The middle latitudes of North America were much more likely than little Norway to see the types of winter storms described by the Norwegian model. But Meisinger seemed not to be discouraged, and he took every opportunity to expound on the new storm model insights of Vilhelm and Jacob Bjerknes in Bergen, Norway. "He still had the enthusiasm of the young man in his studies and fortunately was given considerable leeway in the pursuit of his studies in the Weather Bureau," wrote Ivan Ray Tannehill, who also became a bureau meteorologist. "The inevitable disappointments and numerous complexities that dampen the ardor of the older men in the weather service had apparently not yet affected him."

Analyzing a large and powerful storm that crossed the United States in the middle of February 1919, Meisinger observed a "striking accord" between the observations and the new Norwegian cyclone model. "From the time the storm freed itself from the topographical hindrances of the Rocky Mountains the distribution of winds and precipitation during its eastward march conformed perfectly with the mechanical outline of Bjerknes," he wrote in the *Monthly Weather Review,* October 1920.

In a 1921 article in *Science* magazine reporting on the International Balloon Race of 1920 in Birmingham, Alabama, Meisinger noted that the competition had been won by a Belgian meteorologist who had followed the advice of Weather Bureau observer C. George Andrus, another young ballooning enthusiast. Predicting the arrival of a storm during the second night, Andrus had urged the competitors to avoid westerly progress as much as possible until then. The winner caught the strong upper winds above the *steering line,* or warm front, of the advancing cyclone the following evening and quickly ended up in Vermont. What Meisinger found especially interesting was that Andrus had analyzed his wind observations according to the new cyclone model developed by Jacob Bjerknes. Meisinger wrote that the "charm of the Bjerknesian interpretation is that it enables one to get a more satisfactory three-dimensional picture of the processes taking place in *Highs* and *Lows* than has been usual . . . it was meteorology that won the race."

A tireless worker and a prolific writer throughout the time he was working at the bureau and attending graduate courses at George Washington University, Meisinger published regularly in scientific journals and popular new aeronautics magazines. He translated important foreign scientific papers, including documents describing the German military weather services that were captured after the Great War, and published them in the *Monthly Weather Review* of December 1919. Meanwhile, he was instrumental in helping to found the American Meteorological Society, devoting many hours to organizing meetings, writing reports, and presenting papers. And somehow he found time in 1921 to marry Miss Helen B. Hilton of Lincoln, Nebraska.

Since first coming to the bureau, Meisinger had frequently advocated the use of free manned balloons to expand knowledge of the upper air. In early 1924, an opportunity for an ambitious series of scientific ascents finally presented itself. Ever since his first experimental constant-level balloon flight from Fort Omaha in the spring of 1919, Meisinger wrote in the *Monthly Weather Review* that he "cherished with undiminished enthusiasm the desire to make a series of such flights, carefully planned to take advantage of selected weather types and to utilize to the fullest extent synchronous observations at the surface and in the free air. The time seems now to be propitious for this effort."

Meisinger described two main scientific goals, both relying on constant-level flight and synchronous observations by Weather Bureau observers at stations below the prospective flight paths of the balloon. First was the validation of his work on deriving the direction and strength of winds in the upper air from surface observations. "We now have at our disposal the machinery for making free-air pressure maps, and the one big problem in connection with them is to learn to interpret them," Meisinger wrote in a bureau memorandum. "Can we have confidence in them in deriving the 'life history' of free-air currents? The free balloon (manned) furnishes the only means of obtaining an observation upon the path of air moving in response to given pressure gradients." Using Meisinger's methods, observers on the ground would calculate the pressure pattern at the elevation of the balloon, and the results would be compared with the values obtained by Meisinger during the flights.

His other goal was to develop more fully information about the thermal and wind structure of the atmosphere above cyclones, a purpose that would require the balloonists to cross through their turbulent fronts. Citing Bjerknes and others, Meisinger wrote that recent studies of the mechanism of storms "have served to emphasize the great and immediate necessity for observations as to the nature of air trajectories at various levels in cyclones and anti-cyclones."

Meisinger originally hoped for as many as 15 flights in the spring of 1924. Beginning on April 1, the flights launched from Scott Field, in southern Illinois, where the Army Air Service Balloon and Airship School had opened in 1922. Spring had been chosen because during this season in the Midwest, it was thought that the balloonists could explore the motions above storms without the distractions and dangers of the convective currents of thunderstorms. As it happened, however, the spring of 1924 turned out to be unusually turbulent; it was as unstable as most summer patterns in the region.

Details of these last flights were reported by bureau meteorologist Vincent E. Jakl in the *Monthly Weather Review* of March 1925 and by John M. Lewis and Charles B. Moore in the *Bulletin of the American Meteorological Society* in January and February 1995. Meisinger and his pilot, Lieutenant James T. Neely of the Army Air Service, found themselves whipped around through a series of treacherously wild and wearisome rides in the stormy atmosphere throughout April and May. Maintaining constant levels of altitude had proven especially difficult. Tired and disappointed, Meisinger realized that the undertaking was not going to be as productive of scientific data as he had hoped. The tenth flight, he decided, would be the last of the series. Unfavorable weather delayed this last mission three or four days, and Meisinger and Neely waited impatiently. Finally, on June 2, despite continuing concern for thunderstorms again that night, they lifted off from Scott Field at 4 P.M., hoping to remain aloft for 48 hours.

As the sun set about three hours later, the loss of its radiation caused the balloon to descend to about 1,000 feet. Neely threw off ballast, a 30-pound bag of sand, and the balloon rose to 5,200 feet, although almost immediately it began a more rapid descent. At 1,000 feet, more ballast was discarded, and again the balloon climbed, this time to an altitude of 5,900 feet into a cloud over Macon, Illinois. The cloud was part of a developing thunderstorm, and the balloon was caught in its downdraft and fell in a rapid descent. Neely threw off more sand at about 1,000 feet, sending the balloon back into the clouds. This time they climbed to 7,440 feet in an effort to get above the storm, but the balloon entered another powerful downdraft that plunged it into a rapid, uncontrolled fall. The basket hit the ground with a smash in a cornfield near Milmine, Illinois. The shock dislodged four more ballast bags, and the balloon rebounded with a yank back into the sky.

At 1,200 feet, at 11:14 P.M., a single bolt of lightning struck the balloon. It passed through Meisinger, killing him instantly. It ignited hydrogen leaking from the appendix at the bottom of the balloon, causing a fire in the rigging. Meisinger's body fell to the bottom of the basket. Neely, badly burned, grabbed a parachute and leaped from the basket, but the canopy opened before he was able to attach it to his harness. The 1,200-foot fall killed him instantly. Soon the fire burned through the rigging and the basket carrying Meisinger's body fell away into a pasture, where it was found the next morning by a farmer looking for his horses. Meisinger and Neely both were 29 years old.

"Despite his short term, Meisinger must be considered one of the most prominent figures in American meteorology," Jerome Namias, another great meteorologist, would observe many years later. United States Weather Bureau officials paid tribute to his courage and the value of his contributions to the science. But Weather Bureau forecasters did not adopt Meisinger's method of deriving upper-level winds from surface observations, and soon after his death the agency abandoned his whole line of research. To outstanding young meteorologists, including Namais in 1938, the American Meteorological Society presents each year the Clarence LeRoy Meisinger Award.

PART IV

Together at the Front

•

Nothing on the horizon at the beginning of the twentieth century encouraged the idea that useful advances in weather forecasting were on the way. The fracture in meteorology was real: studying the atmosphere was a science, and predicting weather was an art. Scientists and forecasters were not talking to one another because they had nothing to say.

Almost out of nowhere came one voice calling for the "unification of meteorology." While the manifesto proposed by Vilhelm Bjerknes provoked interest among the wider community, to most weathermen on both sides of the divide, the Norwegian physicist was dreaming.

What changed everything was war. The imperatives of modern warfare put a new face on weather information of all kinds, and enormous scientific and financial resources very quickly were marshaled toward its development. Meteorology in the twentieth century came to be seen as vital to the survival of nations.

18

Vilhelm Bjerknes

The Bergen Schoolmaster

•

WEATHER SCIENCE INSINUATED itself into the life of Vilhelm Bjerknes while he was making other plans. Through the last decade of the nineteenth century—his formative years as a young Norwegian scientist—his focus was mathematical and mechanical, the areas his famous father was pursuing in an effort to explain Newtonian physics by the presence of a cosmic ether. Carl A. Bjerknes ultimately failed to achieve this goal, and while the worldview it proposed lost favor with leading physicists, Vilhelm held the course with the devotion of a son to a father. Along the way, he developed important mathematical principles in the field of hydrodynamics, the motions of fluids, and colleagues at the University of Stockholm quickly saw their value to the fluid realm of geophysics. To Vilhelm, this application to these "alien sciences" was an interesting argument for the larger case he was trying to make, and he drove himself to the edge of mental exhaustion trying to complete his brilliant father's work. But Vilhelm had watched Carl Bjerknes grow ever more isolated in his singular scientific search, and by the turn of the century he knew that he had some painful decisions to make about his own future as a physicist. Finally, with a sense of disappointment, he lay the great work aside and followed the advice of colleagues to turn the power of his circulation theorem to the realms of the atmosphere and the ocean.

Once accomplished, this unexpected and unwelcome twist in his plans seemed to mark a turning point not only in the subject of Bjerknes's research, but in his attitude toward change itself. He was still his father's son, but Carl Bjerknes's solitary and obsessive journey down a single path of theory was not going to be Vilhelm's way. He would learn not only to survive change—the turmoil of World War I, economic depression, and food shortages—but to employ these circumstances to the advantage of his science. He would become adept not only at organizing and pursuing his research goals, but in spreading the influence of his science around the world. More than once he would pause to reflect about life being so "fateful." In fact, he had taken fate into his own hands better than most scientists of the time. Before he and the young scientists he employed were done with the subject, they would be seen as important agents of change, even scientific revolutionaries, and the forecasting practice and the physical theory of meteorology would never be the same. The research goals he laid down for himself and others would set the agenda for 100 years of weather science. Vilhelm Bjerknes was the father of modern meteorology.

In 1904, in a German meteorological journal, *Meteorologische Zeitschrift,* Bjerknes articulated a set of disarmingly plain ideas that defined his long-term goal: to render meteorology an exact science. "If it is true," he wrote, "as every scientist believes, that subsequent atmospheric states develop from the preceding ones according to physical law, then it is apparent that the necessary and sufficient conditions for the rational solution of forecasting problems are the following: 1. A sufficiently accurate knowledge of the state of the atmosphere at the initial time. 2. A sufficiently accurate knowledge of the laws according to which one state of the atmosphere develops from another."

These ideas, straight from the exact sciences of physics and astronomy, must have sounded perfectly sensible to people outside meteorology, especially to users of forecasts. To weathermen of all stripes at the turn of the century, however, they were revolutionary and largely unwelcome. Bjerknes must have been dreaming. Meteorology was as fractured as ever. Scientists and forecasters were very different types of people doing very different kinds of work. Among the theorists and the empirical forecasters, this divide between the branches of meteorology had always existed and was perfectly natural. These people had nothing to say to one another. Besides, in 1904, neither component of his scheme could be said to be in place.

While important advances had been made, the nineteenth century's contributions to theory were a disorganized assortment of principles and rules that had been handed down from the works of astronomers, physicists, and mathematicians who seemed always to have had something more rewarding to do. Weather science had little standing as an academic discipline and no institutional strength. Universities did not offer degrees in the subject, and there were few journals of note. Communication among the few far-flung researchers was so haphazard that often they found themselves reinventing each other's work. For every purpose, observational data was spotty and unreliable. The best scientists were contemplating the machinations of idealized atmospheres that bore little relation to what surrounds Earth, or any other planet, and the results of their research were beyond the interest or understanding of meteorology's practical men.

Forecasters at this time were practicing an already moribund art, one nearly immaculate of science. The government agency men in charge of the forecasting services had spent 30 years extending the geographical reach of their systems without extending their intellectual grasp. While the science of weather was still young as a newborn, already its application was in the hands of intransigent old men. The high expectations that had come in the first days of storm-warning services had long since faded. Weather forecasting was not very reliable, everyone knew, and yet few people—its practitioners least of all—seemed to expect it to get much better. While its scientific principles were amazingly few, to the forecasting men they were seen as perfectly adequate to the state of the art.

This work was not so much about physics as geometry—plotting shapes on maps according to the latest telegraphic observations. The formation of storms, and their potential for growth or decay, was beyond the scope of this process. The concept was vintage 1870s and as simple as could be: weather traveled west to east, and a storm was defined by the presence of low barometric pressure. On maps they drew *isobars,* lines connecting dots marking the same surface air pressure, and then smoothed their contours. Comparison with previous maps allowed forecasters to trace the movement of a center of low pressure, its path and pace. Its next movements would then be approximately advanced, according to a process that could not easily be reduced to description. There were charts that showed preferred storm tracks. There were generalizations about seasonal changes and a bewildering assortment of rules of thumb that were dubious and even contradictory. In the United States, Weather Bureau forecasters were bred from within, trained not in the sciences but on the job, in the old apprenticeship model, spending years watching the master around the map. This system had no place for some new theory of scientific forecasting, and for decades the old men of the U.S. Weather Bureau would ignore the revolution sweeping their field.

Not everyone in the United States was of the same mind as the men in the Weather Bureau. As it happened, ironically, U.S. financial support was crucial to the early nurturing and ultimate scientific success of the new Bergen school of meteorology that Vilhelm Bjerknes eventually established in Norway. In 1905, Bjerknes, a 43-year-old professor at the University of Stockholm, brought his meteorological manifesto to the United States, accepting invitations to lecture at Columbia University in New York and at the Carnegie Institution in Washington, D.C. He explained his circulation theorem and how known physical principles of hydrodynamics and thermodynamics could be applied to the atmosphere. He called for the unification of meteorology and the need to bring physical theory to bear on weather forecasting; and he expressed the ambitious conviction that in theory and practice, the science of the weather could be made exact. Most radically, as Bjerknes put the case, the measure of the science's exactitude would be its ability to foretell accurately the change in the weather from one time to another. Carnegie president Robert S. Woodward was enthusiastic and became a lifelong ally. Bjerknes was named a Carnegie research associate and granted an annual stipend that would allow him to employ young research assistants wherever he went. This Carnegie Institution support lasted the rest of Bjerknes's professional life—until 1941 and the outbreak of World War II. Looking back over the century of science it spawned, Arnt Eliassen, himself an eminent Norwegian geophysicist, recently wrote, "The money could hardly have found a better use."

These Carnegie assistants would become leading geophysicists of the twentieth century, formulating principles about the circulation of ocean currents, the character of storms and their life cycles, the structure of the atmosphere, and other important theory. Moreover, by fanning out from Bergen and spreading the word at a vital moment in history, they would completely reform the practice of weather forecasting around the world, improving its accuracy and extending its reach and usefulness. The work of these men was critical to the success of military and civilian aviation.

Bjerknes was a gentle and thoughtful man, a brilliant thinker and a creative scientific leader who had a rare ability to attract and inspire bright young students. In choosing research assistants, he had a Midas touch that was enriched by his ability to strike the right combination of guidance and freedom. "Wherever he was," noted Eliassen, "in Stockholm, Oslo, Leipzig, or Bergen, he created a fertile milieu around himself." He was a careful listener who was quick to lend insight, encouragement, and moral support. "An intellectual type, Vilhelm Bjerknes was distinctively visual; whenever possible, he would give geometrical interpretations to his equations." Invoking the memory of another brilliant twentieth-century meteorologist, Eliassen wrote: "Jule Charney once said that he thought

Vilhelm Bjerknes in trying to understand atmospheric dynamics would imagine that he himself was an air particle, trying to decide where to go."

Among his young Swedish students in Stockholm was Vagn Walfrid Ekman, whose assignments from Bjerknes led to important discoveries about the ocean and the atmosphere. The research stemmed from discussions between Bjerknes and Fridtjof Nansen, a Norwegian explorer who, on ship and skis, spent the years 1893 to 1896 crossing the Arctic Ocean. Off the coast of Siberia, Nansen's vessel was suddenly stopped dead in the sea. Studies by Ekman confirmed Bjerknes's speculation that the mysterious "dead-water" phenomenon was caused by internal "gravity waves" at the intersection of fresh surface water and more saline water below. Another Ekman investigation into a puzzle from the expedition explained mathematically how Earth's rotation caused ocean currents to drift to the right of the winds. What came to be known as the *Ekman spiral* also causes winds to drift (to the right in the Northern Hemisphere) as they encounter the effects of friction in the lower atmosphere.

In 1907, Bjerknes returned to Norway to accept a professorship at the University of Christiania (later Oslo), where he continued his theoretical studies of the atmosphere. But times were changing. The conquest of the air represented first by the advent of rigid gas-filled airships and then by airplanes was quickly recognized by Bjerknes as a powerful boost for meteorology. Regular flights could mean regular soundings of the upper air, vastly improving prospects for theory and forecasting. At the same time, manned flight immediately made new demands on the ability of the science to produce more accurate and more detailed weather information. Among his young Norwegian research assistants was Harald Ulrik Sverdrup, who would go on to become a famous oceanographer.

In 1912, Bjerknes accepted an invitation to become the founding director of a new geophysical institute at the University of Leipzig. The city was becoming an important center for Zeppelin airships, which relied heavily on weather forecasting, and war was on the horizon. Bjerknes found German military planners and aeronautical officials to be more responsive and careful listeners to his discussions about the study of the upper air and how best to make observations than audiences in Norway and elsewhere in Europe. Unlike Britain and France, Germany recognized even before 1914 the critical role weather services would play in the successful use of the emerging technologies of a modern military: aviation, gas warfare, and long-range artillery.

In his inaugural lecture at Leipzig in 1913, reprinted in the January 1914 *Monthly Weather Review,* Bjerknes reviewed the advances in theory that had inspired his ambitions for meteorology. Now the equations were in place to characterize mathematically the principle conditions of the atmosphere. He described how the recent advent of aeronautical observatories across Europe finally was producing regular readings of conditions in the "free air" of the upper atmosphere that were so crucial to his plan. And now, with this information in hand, he declared,

> [A] mighty problem looms before us and we can no longer disregard it. We must apply the equations of theoretical physics not to ideal cases only, but to the actual existing atmospheric conditions as they are revealed by modern observations. These equations contain the laws according to which subsequent atmospheric conditions develop from those that precede them. It is for us to discover a method of practically utilizing the knowledge contained in the equations. From the conditions revealed by the observations we must learn to compute those that will follow. The problem of accurate pre-calculation that was solved for astronomy centuries ago must now be attacked in all earnest for meteorology.

Bjerknes's work was well funded and well exposed, and he was supplied with numerous research assistants. In addition to his German assistants, Bjerknes was joined by two new Norwegian Carnegie assistants: his son, Jacob Bjerknes, and Halvor Solberg. Bjerknes reported making modest progress in his effort to use hydrodynamic equations to forecast the behavior of winds, but that progress would not last. With the outbreak of World War I in 1914, working and living conditions in Leipzig grew increasingly more difficult. One after another of his German research assistants was forced to leave his laboratory, having been called to the front, where many of them were killed. Food and fuel became more and more scarce. Finally, in 1917, Bjerknes left Leipzig with his son and Solberg, returning to Norway to found a geophysical institute in Bergen.

In western Norway, Bjerknes's scientific enterprise took another sharp turn. Neutral it may have been, but its noncombatant status had not allowed Norway to escape the effects of the war. The large Norwegian fishing fleet had been cut off from the system of telegraphic storm warnings from Britain, where weather forecasts had become secret military intelligence. With the disruption of normal trade, Norway faced food shortages so severe that they raised the specter of famine. Norway imposed rationing and took emergency measures to boost the output of its farmers. Bjerknes quickly recognized the role that his new science could play in the national emergency and took a proposal directly to the prime minister. He offered Norwegian agriculture "all the support it can receive from meteorology's side." With that pledge, and government support, the academic Geophysical Institute of Bergen transformed itself in 1918 into an operational weather center, and its meteorological theorists became very practical forecasters.

This 1918 initiative led to the most productive period of Bjerknes's long career in science. The dense network of sea and land observations the Bergen scientists established, and the intense physical analyses they applied to the weather maps during the critical growing season that summer, gave rise to the discoveries that would become the hallmarks of the Bergen school of meteorology, which would change the face of weather forecasting around the world. It would not be *dynamic* meteorology—the application of the equations of hydrodynamics and thermodynamics—that would make the first big impression on the science. It would be the new attitude they brought to *synoptic* meteorology, the study of maps displaying simultaneous observations of specific weather systems. They came to weather forecasting with virtually no training, and so were unfettered by the knowledge of all that was not possible. Their fresh eyes were reading maps of observations with the intensity of youth and the intelligence of highly trained geophysicists.

Analyzing the lines of converging winds of a particularly troublesome mid-August storm, Jacob Bjerknes began sketching the details of what became the new model for the structure of a midlatitude cyclone. Others as far back as Luke Howard in the 1820s, Elias Loomis in the 1840s, and Heinrich Dove in the 1850s had caught glimpses of its features. But it was 22-year-old Jacob Bjerknes whose classic paper "On the Structure of Moving Cyclones" depicted for the first time in October 1918 a midlatitude storm as a hook-shaped center of low pressure with two distinct lines of weather separated by a wedge of warm air. Inspired by the war, the battleground terminology of *fronts* would become part of the new lexicon of weather the following year.

Another conceptual breakthrough followed from the other young Norwegian Carnegie assistant, Halvor Solberg, who studied old charts of the North Atlantic and identified "families" of cyclones in different stages of development along a line around the hemisphere that the Bergen school called the *polar front*. A research assistant from Sweden, Tor Bergeron, completed the new cyclone model by recognizing how the faster-moving cold front closes in on the warm front, squeezing off the flow of warm air to the storm's center and causing its decay.

Carl-Gustaf Rossby, the Swede, spent only a short time in Norway, but he became one of the Bergen school's most influential adherents. It would be Rossby who eventually overcame the resistance of the U.S.

Weather Bureau to bringing the new forecasting methods and theoretical tools to the United States.

In 1926, Vilhelm left the Bergen school in the hands of his son and accepted a professorship at the University of Oslo, where he resumed work on his dream of bringing the physical laws of hydrodynamics and thermodynamics to bear on the problems of meteorology—on both weather forecasting and theory. He would not live to see this dream accomplished. He was wholly aware of this circumstance and did not allow it to dissuade him from his goal. As he told his Leipzig audience in 1913, "One does not always aim only at what he expects soon to attain. The effort to steer straight toward a distant, possibly unattainable point, serves, nevertheless, to fix one's course."

Critics who complained that Bjerknes's methods of calculating weather changes take longer than the weather itself were missing his point. "I shall be more than happy if I can carry on the work so far that I am able to predict the weather from day to day after many years of calculation," he said.

> If only the calculation shall agree with the facts, the scientific victory will be won. Meteorology would then have become an exact science, a true physics of the atmosphere. When that point is reached, then the practical results will soon develop.
>
> It may require many years to bore a tunnel through a mountain. Many a laborer may not live to see the cut finished. Nevertheless this will not prevent later comers from riding through the tunnel at express-train speed.

As Vilhelm Bjerknes predicted, it would be many years before the express trains would be roaring through the tunnel, although it would not be as long as anyone in 1913 would have guessed. Certainly he would not have predicted that it would be only a matter of months before Lewis Fry Richardson would begin so laboriously to cut the first holes in the mountain.

19

Lewis Fry Richardson

The Forecasting Factory

•

AGAINST ALL PROBABILITY, the first attempt to forecast weather by solving the equations that describe the physics of the atmosphere came heroically amid the battles of World War I. That an Englishman in his mid-30s made the effort in the second decade of the century is a wonder of science and human ingenuity. So complex were the problems, so tedious were the solutions, and so weak were the observational data at the time that not even Vilhelm Bjerknes was tempted to practice what he preached. That anyone could do it so well at the time was practically miraculous. Reviewing the results 70 years later, and marveling at the "buoyancy of spirit" it must have required, one meteorologist described the mathematics as "one of the most remarkable and prodigious calculational feats ever accomplished." Ingeniously, that Englishman conceived and operated a "computer model" of the atmosphere before the age of computers. So how in the world did Lewis Fry Richardson carry out such a mission while driving an ambulance through the muddy annihilation of The Great War?

For two years, beginning in the fall of 1916, Richardson served behind the trenches on the other side of the Western Front from the German researchers who had been working with Vilhelm Bjerknes at the Geophysics Institute at the University of Leipzig. Ten of Bjerknes's young German scientists had been called away from his laboratory, crippling his research program. Five of them died at the front, one apparently within 50 miles of Richardson's position near the fourteenth French Army, which was dug in on the plain of Champagne, between the Germans and

Paris. These were the circumstances that prompted Bjerknes, in the summer of 1917, to return to Norway, where he shifted his focus from his long-range theoretical pursuit to the more immediate demands of forecasting weather. So it happened in 1917 that Richardson, behind the French trenches, single-handedly was engaged in the very enterprise that Bjerknes, behind the German trenches, was being forced to abandon.

A Quaker and a pacifist, Richardson was part of the Friends Ambulance Unit that had been formed in England to give conscientious objectors a way to perform their public duty in a noncombatant role in the national emergency. But the 36-year-old scientist was not subject to military draft at the time, and probably never would have been. He had joined the war voluntarily after giving up what may have been the best job he would ever have. The dictates of "an inner programme" were at work and nothing, it seemed, was going to keep this conscientious objector from the battlefields of World War I.

Richardson was born in 1881, the youngest of five boys and two girls in a prosperous Quaker family at Newcastle upon Tyne, an industrial port in the north of England. His father was David Richardson, a tanner, and his mother, Catherine Fry, came from a family of grain merchants. He attended a private Quaker boarding school in York, then Durham College of Science in Newcastle, and then King's College in Cambridge. Why did he train in physics and other sciences, in which he had no particular focus, rather than mathematics, for which he had an uncommonly strong aptitude and lifelong passion? All his life he would frame problems and solutions in challenging mathematical terms, a predilection that did not always serve him well.

Richardson spent 10 years in a variety of short-term scientific jobs, including stints at Sunbeam Lamp Company and National Peat Industries Limited. In 1909, he married Dorothy Garnett, the sister of a Cambridge colleague, who came from a scientific family. Unable to have children—Dorothy would suffer seven miscarriages—the Richardsons adopted two boys and a girl. Dorothy would be the mainstay of the family; she would run the household, compensating for her husband's impractical nature, and give emotional, moral, and intellectual support to a man who worked most often in solitude and isolation.

"I hardly know what loneliness feels like," he wrote. "When solitary, I am usually serene; when in a crowd, I am often embarrassed." This "temperamental peculiarity" he saw as both a strength and a weakness. "I have always had an inner programme, often secret, often frustrated, and in retrospect sometimes silly, sometimes wise." His unconventional approach gave his science originality, but often he faced resistance to his methods or views. To the British geophysicist Geoffrey I. Taylor, Richardson was "a very interesting and original character who seldom thought on the same lines as his contemporaries and often was not understood by them."

Richardson's first important scientific paper, published by the Royal Society of London in 1910, was titled "The Approximate Arithmetical Solution by Finite Differences of Physical Problems Involving Differential Equations, with an Application to the Stresses in a Masonry Dam." In his industrial jobs, he had formulated some innovative ways to solve differential equations, which express in the symbolic language of calculus the relationships of things that change. He was looking for ways to put his techniques to good use, believing, as he wrote in 1908, that "the greatest stimulus of scientific discovery are its practical applications." It would be five years before Richardson first settled on meteorology as a physical problem in need of his approximate arithmetical solution by finite differences.

The idea came to him in 1913. He had just been appointed superintendent of Eskdalemuir Observatory, a geomagnetic and meteorological research facility in Scotland that was operated by the British Meteorological Office and the Royal Society of London. This comfortable and secure position certainly was his best in the 10 years since he had left the university. It provided a salary of £350, a handsome sum at the time; a house on the grounds; a staff of assistants; and exactly the kind of remote, solitary environment that he always craved.

The idea first came to him as a fantasy—a vision of a great "forecast factory" that would be as good a pictorial representation of the work done by modern computer models of numerical weather prediction as has ever been written. In Richardson's day, in Richardson's mind, a "computer" was a person. He saw "a myriad computers" working in a large hall that was like a theater, except that it was completely encircled with galleries, without a stage. "The walls of this chamber are painted to form a map of the globe," he wrote. All around the map were stationed the individual computers, each responsible for calculating changes in his one small unit of the global atmosphere. From the floor of the hall there rose halfway to the ceiling a pillar that supported a pulpit, where a man in charge, "like a conductor of an orchestra," kept the computers working their slide rules and calculating machines in time with one another. "Four senior clerks in the central pulpit are collecting the future weather as fast as it is being computed, and despatching it by pneumatic carrier to a quiet room," Richardson wrote. "There it will be coded and telephoned to the radio transmitting station."

To anyone else, this might have been an idle day dream. To Richardson, there was nothing idle about it. It was full of meaning, the product of a creative process he called "intentionally guided dreaming," and he took it seriously. The forecast factory represented to Richardson not only his vision of the future of weather prediction, but a graphic depiction of the ambitious program of mathematical research and arduous calculation he was about to embark upon.

Richardson described this fantasy to his boss, Napier Shaw, head of the British Meteorological Office, who then wrote to Vilhelm Bjerknes asking for help in getting copies of Bjerknes's two recent books on the subject of the dynamics of the atmosphere to the interesting new man he had recently posted to rural Scotland. Shaw told Bjerknes: "He presented me the other day with what he called a dream of a palace at the Hague in which 500 computers were taking down the observations for all parts of the world read out from a conductor's box in the middle of a great theatre, each computer dealing with the observations for one compartment. I explained to him that you had already set out upon the programme and recommended him to digest what you had already done."

In the preface of his book *Weather Prediction by Numerical Process,* Richardson acknowledged his debt. "The extensive researches of V. Bjerknes and his School are pervaded by the idea of using the differential equations for all that they are worth," he wrote. "I read his volumes on *Statics* and *Kinematics* soon after beginning the present study, and they have exercised a considerable influence throughout it."

At Eskdalemuir, between 1913 and 1916, and between his official administrative, research, and observational duties at the observatory, Richardson completed the first draft of his book. He had taken the fundamental equations that were known to describe the behavior of the atmosphere and transformed them in ways that allowed him to solve them by the approximate methods he had described in 1910. Other problems required him to work out his own equations from basic physical principles. "The fundamental idea is that atmospheric pressures, velocities, etc., should be expressed as numbers, and should be tabulated at certain latitudes, longitudes and heights, so as to give a general account of the state of the atmosphere at any instant, over an extended region, up to a height of say 20 kilometres," he wrote. He had worked out "an arithmetical method of operating upon these tabulated numbers, so as to obtain a new table representing approximately the subsequent state of the atmosphere after a brief interval of time." The set of "computing forms" he worked out correspond to the software program instructions of a modern Numerical Weather Prediction computer model.

Napier Shaw presented the first draft to the Royal Society, which pledged £100 for its publication, but Richardson decided that he was not ready to see the book in print. He had developed the mathematical methods and organized the concepts illustrated in his forecast factory fantasy, but he had not operated the factory. He had not put his formulations to the test. He had not attempted a forecast. This incredibly difficult job of sustained calculation probably could have been most comfortably accomplished in the peace and quiet of the remote Scottish lowlands of the Eskdalemuir Observatory.

But World War I had been weighing heavily on Richardson for two years, despite the fact that as a practical matter there was no need for him to be there in the thick of it in 1916. His position as superintendent of the Eskdalemuir Observatory would have been considered of sufficient national importance to exempt him from military conscription. In any case, as a married man, now 35 years old, he would not have been drafted in 1916. But Richardson had his own reasons for putting himself in harm's way. A profoundly committed pacifist, Richardson would devote many years of his professional life to conflict resolution and the sociology of war, and he felt compelled to endure the experience. Much later in his life, he would describe the personal emotional struggle he had been feeling in Eskdalemuir since the outbreak of the war: "In August 1914 I was torn between an intense curiosity to see war at close quarters, an intense objection to killing people, both mixed with ideas of public duty, and doubt as to whether I could endure danger."

So he resigned his well-paid, secure scientific position in May 1916 and took the draft of his manuscript and his enormous calculation job with him to war. "The manuscript was revised and the detailed example . . . was worked out in France in the intervals of transporting wounded in 1916–1918," he wrote in his book, giving only the barest outline of the conditions he endured. "My office was a heap of hay in a cold rest billet." Life with the Fourteenth French Army was difficult and dangerous and occasionally chaotic, and Richardson describes how at one point the entire enterprise almost came to nothing: "During the battle of Champagne in April 1917 the working copy was sent to the rear, where it became lost, to be rediscovered some months later under a heap of coal."

For his calculations, Richardson used published data sets provided by Vilhelm Bjerknes for May 20, 1910, which had been an "international balloon day" organized by observatories throughout Europe for taking special measurements of the upper air. He would test his method by manipulating the data to "pre-calculate" the changes six hours into the future, and then he would compare his result with the actual changes recorded by observations that were taken later that day.

This famous, first-of-its-kind "forecast" he so patiently hand-calculated under such difficult conditions failed spectacularly, producing what Richardson himself called a "glaring error." His method generated changes in air pressure over a six-hour period that were entirely unrealistic—100 times the values of changes that were actually observed six hours later that day in 1910. Richardson blamed the error on bad wind data. A modern evaluation of Richardson's forecast points to technical problems in the process known among modern computer modelers as *initialization* as the chief difficulty with the output of his equations. That Richardson would eventually proceed with publication of his failed forecast speaks to the strength of his character as a scientist. Moreover, Richardson understood, as many others did not, that the importance of his work was the methods he had devised, and the fact that they did not work perfectly the first time they were tried was beside the point.

His noncombatant ambulance service gave Richardson a certain sense of distance from the worst of the war—from the killing—as it was clearly intended to on moral grounds. And certainly his elaborate mathematical and other scientific pursuits offered him more diversion than the free-time occupations of most soldiers. But he and the other 55 men who drove the 22 ambulances that comprised the Section Sanitaire Anglaise 13 were not spared the traumatizing psychological impact of this first use of the weapons of mass destruction. More than once they came under heavy artillery shelling, especially during the battle of Champagne in April 1917, as they shuttled wounded from a battleground aid station to a triage unit in a nearby village. Richardson, who returned to England in 1919, came away from the experience with the same psychological infirmities of many soldiers returning from combat. His son, Stephen, would remember the terrorized look of shell shock when he startled his father, and long afterward the sounds of recurring nightmares. Richardson wrote prolifically about wars, characteristically quantifying their causes and cures with his complex equations. But in common with many a weary veteran, he seems to have written hardly a word about his personal experiences at the front.

Weather Prediction by Numerical Process was not published until 1922, six years after he had completed his first draft near the end of his tenure as superintendent at Eskdalemuir. As he described it, "the legacy of war" had delayed its publication after his return. Shaw had welcomed Richardson back as a researcher for the Meteorological Office, assigning him duty as a researcher at the Benson Observatory under William H.

Dines, a pioneering researcher of the upper atmosphere and radiation. Dines was 25 years older, but the two men enjoyed talents and temperaments in common—passions for science, originality in devising instruments and experiments, and insular, retiring dispositions. Richardson rewrote the manuscript in 1920 and 1921, acknowledging help from Dines with certain refinements.

Despite the influence of Vilhelm Bjerknes and the guidance from Dines and others, *Weather Prediction by Numerical Process* was not going to be mistaken for the work of anyone but Lewis Fry Richardson. The iconoclastic mix of intensely complex mathematics and playful whimsy is all Richardson. Some of the most difficult calculations ever attempted are followed by the fantasy of his forecast factory. In a section on turbulence and eddy motion in the formation of cumulus clouds, Richardson observes that "big whirls have little whirls that feed on their velocity, and little whirls have lesser whirls and so on to viscosity—in a molecular sense."

He made no apology for the complexity of his subject or the demands it made on scientific colleagues who were less adept than he at the nuances of differential calculus. "The scheme is complicated because the atmosphere is complicated," Richardson wrote. Weather is a messy process of unbridled forces tearing at the edge of chaos, and the realities of geophysical fluid dynamics do not yield to the kind of simplicity attained in other physical realms. In a note to himself, Richardson observed how Einstein and other theorists had intuitively preferred simplicity and elegance in their formulations, remarking that they had been "brilliantly successful" in their spheres. "If they would condescend to attend to meteorology the subject might be greatly enriched," he wrote. "But I suspect that they would have to abandon the idea that truth is really simple."

He called readers' attention to the computing forms, which were intended to reduce the volume of calculation. Richardson estimated that 64,000 individual computers would be required to keep pace with weather around the globe, although perhaps his whimsy got the better of him. Recent recalculations by meteorologist Peter Lynch of the Irish weather service puts the figure at 204,800. Whatever the number, Richardson made his point: "Perhaps some day in the dim future it will be possible to advance the computations faster than the weather advances and at a cost less than the saving to mankind due to the information gained. But that is a dream."

In 1922, the book was politely reviewed, even hailed as an important achievement, but it was not much understood. Not many scientists engaged in meteorology in those days were entirely capable of following

Richardson's complex mathematics. And before the advent of electronic computers, almost no one besides Vilhelm Bjerknes himself took seriously his dream that the day would come when science's ability to compute the physics of the atmosphere would outpace changes in the weather. The failure of Richardson's forecast may have discouraged others from attempting such an arduous computational approach, especially in light of the more obviously practical discoveries that were coming from the Bergen school at the time. But even in the most astute hands, and even with the best of data, the method he proposed was just too intrinsically difficult and time consuming to be seen as a practical solution to forecasting weather.

Richardson must have been disappointed that virtually no one among the dynamical meteorologists followed up his Numerical Weather Prediction methods with further experimentation. But even before the book was published, circumstances beyond the science were conspiring to drive Lewis Fry Richardson away from the science of weather. After World War I, operation of the British Meteorological Office was transferred to the Air Ministry, signaling a new conflict between his scientific hopes and his pacifist conscience. In the summer of 1920, he severed his association with the agency. A later director of the Meteorological Office would describe this decision as "one of the tragedies in the history of meteorology." Richardson continued to work in the science a few more years, but the end was near. In a tribute to her husband published shortly after his death in 1953, Dorothy Richardson recalled "a time of heartbreak when those most interested in his 'upper air' researches proved to be the 'poison gas' experts. Lewis stopped his meteorological researches, destroying such as had not been published. What this cost him none will ever know!" Richardson devoted most of the rest of his years to writing pursuits outside of meteorology.

"Despite the understandably cautious initial reaction," wrote Lynch, "Richardson's brilliant and prescient ideas are now universally recognized among meteorologists and his work is the foundation upon which modern forecasting is built." He did not live long enough to see Numerical Weather Prediction become a reality, but he lived long enough to see it coming. In 1952, Jule Charney, another great meteorologist, sent Richardson several of his journal papers describing the progress of computer modeling of the atmosphere. The aging scientist congratulated Charney on the "enormous scientific advance on the single, and quite wrong, result in which Richardson ended."

Richardson died on September 30, 1953, at the age of 71, at his home in Kilmun, Scotland, quietly, in his sleep, of a heart attack.

20

Jacob Bjerknes

From Polar Front to El Niño

•

AFTER LEAVING LEIPZIG for Bergen in 1917, young Jacob Bjerknes watched his father's visionary approach to the dynamics of the atmosphere give way to more urgent needs in wartime Norway. Like Lewis Fry Richardson, Vilhelm Bjerknes had no doubt that the mysteries of weather would yield to the laws of physics eventually, but faith in the theory was no answer to the looming threat of famine in the fatherland. A precarious isolation had come as the price of neutrality. Cut off by the war from land and sea observations from Iceland and Britain, farmers and fishermen were desperate for a new source of weather information to help them meet the national food emergency. Just 12 years after becoming independent of Sweden in 1905, Norway faced a crisis of self-sufficiency, and Vilhelm Bjerknes understood his patriotic duty.

But what exactly did the three physical theorists from Leipzig—Vilhelm and Jacob Bjerknes, and the other Carnegie assistant, the young Swede Halvor Solberg—know about practical, day-to-day weather forecasting? While the German military had looked to the Geophysics Institute for help with specific problems of battlefield prediction, especially of the upper air, daily forecasting had not been part of the program. The new team intended to turn this to their advantage. They had no intention of mimicking the unreliable old ways of the empirical school: tracing contours of surface air pressure across a map, applying rules of thumb and intuition to divine their relation to the next day's weather. For the first time, the hoary problem of weather prediction was going to be attacked head-on by highly motivated physicists. While they would not be pursuing

Vilhelm's research goal of precalculating the weather according to equations that describe the laws of physics, they had every expectation that they could devise a scientific approach that improved on the state of the art. Little did they know what Jacob so quickly would make of the occasion.

Jacob brought with him one particularly promising line of research that seemed congenial to both theory and practice. Herbert Petzold, the first doctoral student at the Leipzig Geophysics Institute, had been assigned the problem of studying lines of convergence that appeared on maps showing temperature, pressure, and wind fields at the surface. At the institute, Vilhelm Bjerknes had a theoretical interest in these features. At the same time, the new weapons of mass destruction deployed during the war—long-range artillery, poison gas, and aircraft—were making practical new demands on weather scientists on both sides of the trenches. In Leipzig, German aeronautical officials wanted to know if these lines of convergence were connected to line squalls of thunderstorms that could suddenly rear up and threaten the flights of airships and airplanes. Petzold's early studies seemed to confirm the connection between these lethal threats and mapped streamlines that marked sharp contrasts in pressure and temperature and converging winds. After a year of research, however, the young scientist was called to the front, where in 1916 he was killed at Verdun. Petzold's research was resumed by 18-year-old Jacob when he joined his father's laboratory in 1916. In a German meteorological journal that year, Jacob's first scientific paper identified convergence lines as a frontal boundary where cold air is intruding, pushing warmer air up ahead of it, marking "bandformed regions of overcast, ordinary rain, or showers and frontal storms," depending on their strength.

Food rationing was imposed in Norway, and by the spring of 1918 food supplies were running dangerously low. With government financing and strong support from the Norwegian navy, Vilhelm Bjerknes visited every likely outpost in western Norway, setting up a unique network of weather observation stations along the coast and on the islands in the North Sea. What Norway had lost in range of observations from Iceland and Britain, the Bergen team hoped to compensate for with density of local data. Where there had been 9 stations, now there were 90. For the crop-growing season in the summer of 1918, Norway's farmers would have a weather forecasting service like no other in the world. And not so incidentally, the physicists of Bergen would have the most precisely detailed descriptions yet of wind fields and other surface conditions that marked the approach and passage of midlatitude cyclones.

To Jacob, who was responsible for daily forecasting, the network of tightly spaced observations was like looking at the atmosphere under a microscope. Every morning all summer long, a new slide of data was delivered and quickly became the object of intense investigation. And every day the dimensions of the storms that brought the summer rains to western Norway came into slightly sharper focus. Bringing form to the data, Jacob Bjerknes was doing something that all his life he would do better than anybody. From the data and from the maps, the creative intelligence of the young physicist constructed a strikingly new way of seeing the storms of the middle latitudes.

The long lines of convergence were not incidental to the storms, he concluded, but their most important features. By focusing so obsessively on the centers of high and low pressure, conditions they could most easily monitor, traditional forecasters were failing to observe the elements of the storms that were most responsible for their patterns of rain and snowfall. Jacob saw that from the storm's center, two different lines of discontinuity extended radially for many miles. Marked by a broad swath of rain, one line seemed to connect to the storm's center at nearly the same angle as a line describing the storm's eastward progress. This Bjerknes called the "steering line." At a bigger angle radiated a long squall line that was associated with a more confined strip of precipitation. Work by other scientists had been hinting at this structure, but now Jacob made it clear. Middle latitude cyclones were not more or less symmetrical spirals of cold or warm air, as traditional forecasters had been describing them for so long. Rather, they were distinctly different patterns of both warm and cold flows. Between the two long lines of convergence, Jacob identified a characteristic warm sector that extended toward the center of the storm like a tongue surrounded by colder air.

In October 1918, Jacob Bjerknes, now 20 years old, completed a paper, "On the Structure of Moving Cyclones," that would be seen as a landmark in the science of weather. Publication in 1919 of this eight-page article in a journal of the Norwegian Academy of Science and Letters marked the beginning of what would become known as the Bergen school of meteorology. A new era in the science—both the study of the atmosphere and the practice of predicting weather—emerged from Jacob Bjerknes's experience of issuing daily weather forecasts that summer in western Norway.

From Jacob's cyclone model grew a completely new three-dimensional system for seeing and thinking about storms of the middle latitudes and other weather events. The lines of convergence, a term first used to describe wrinkles in the horizontal wind field, evolved in the three-dimensional model into the important concept of *surfaces of discontinuity.* These were zones of rapid transition in winds, temperatures, humidity, and other variables that mark the boundaries between air masses of different character and origin. This is where the business of storms is settled: along three-dimensional seams in the atmosphere where most of the snow and rain falls, where clouds form, and temperatures and winds shift dramatically—at the surface and aloft.

Frustration over the continuing lack of upper-air observations to develop the model's vertical profile prompted another innovation. Just as the Norwegians had sought to compensate for the loss of long-range monitoring by setting up more densely arrayed stations, in 1919 they sought to make up for the lack of direct upper-air measurements with a system of "indirect aerology." Their weather observers would routinely report cloud formations. With the new storm model in hand, careful analysis of cloud forms allowed the Bergen team to infer changes in the vertical structure of the atmosphere as weather swept across Norway.

Between 1918 and 1924, Vilhelm and Jacob Bjerknes, Halvor Solberg, and a cadre of other brilliant young Scandinavian collaborators would construct a whole new conceptual foundation for the science of weather. For the first time, middle latitude cyclones would be defined as waves that form along the polar front in "families" that undergo characteristic and predictable stages of development. Tor Bergeron, a young Swedish student collaborator, would discover how in the later stages of a cyclone the faster-moving cold front overtakes the warm front, forming a stormy region of suspended warm air he called an *occlusion*. Solberg would review old maps showing data across the North Atlantic and discover several cyclones in various stages of development along a distinctive line of discontinuity. The group's work would culminate with Jacob's and Solberg's publication in 1923 of their groundbreaking paper, "Life Cycle of Cyclones and the Polar Front Theory of Atmospheric Circulation."

These concepts, which formed a powerful new analytical approach for both theoretical research and practical prediction, were cast by the Bergen meteorologists in battlefield nomenclature inspired by World War I. Surfaces of discontinuity became *fronts*. In Jacob's model, the squall line came to be called the *cold front* and the steering line the *warm front*. The polar front, which the group saw as a continuous wavy feature encircling the hemisphere, was the main line of battle between tongues of cold air migrating south from the pole and tongues of warm air moving north from the equator. "The warm is victorious to the east of the centre," Vilhelm Bjerknes wrote in a British journal, the *Quarterly Journal of the Royal Meteorological Society*, in 1920. "Here it rises up over the cold, and approaches in this way a step towards it goal, the pole.

The cold air, which is pressed hard, escapes to the west, in order suddenly to make a sharp turn towards the south, and attacks the warm air in the flank: it penetrates under it as cold West wind."

The Bergen meteorologists, meanwhile, had battles of another sort to fight. From the beginning, Vilhelm Bjerknes had called for the unification of meteorology—joining the fractured world of meteorological research and practical weather prediction along common, rigorous new scientific lines. And by the early 1920s, the Bergen meteorologists had formulated a system that allowed theorists and forecasting practitioners to speak the same language. Bjerknes and his young apostles traveled the world to spread the new gospel. Among them was the Swedish student Carl-Gustaf Rossby, who would become one of the most influential meteorologists of the century. "The polar front theory bridged the gap between the two meteorological camps: forecasting and theoreticians," Rossby would write. But from the beginning, the new Bergen concepts faced resistance from the old school of weather forecasters and research meteorologists. There was a long history of distrust in both camps. To forecasters, theorists had offered almost nothing of practical value. To theorists, forecasting remained an "artful" enterprise beneath the realm of science. As Rossby would find, nowhere was the resistance stronger than among the entrenched weather forecasting bureaucrats in the United States, who were the last holdouts against the new concepts. It would be nearly 20 years before the U.S. Weather Bureau would adopt Bergen school forecasting methods, despite their obvious value to the growing aviation industry.

Still serving as head of the weather forecasting service for western Norway, Jacob Bjerknes received a Ph.D. in 1924 on the strength of his work on the cyclone model and polar front meteorology. In 1926, he served as a meteorologist in Iceland in support of Norwegian explorer Roald Amundsen's attempt to fly over the North Pole in a dirigible. In 1928, he married Hedvig Borthen, the daughter of a prominent Bergen ophthalmologist. They had two children. In 1931, Jacob left the weather service in the able hands of his Norwegian colleague Sverre Petterssen and became professor of meteorology at the Geophysical Institute of Bergen. With upper-air data finally available, he focused his brilliant analytical skills on the development of waves in the upper atmosphere and their relation to midlatitude cyclones, producing, in 1937, another seminal paper on the subject.

In July 1939, Bjerknes and his young family traveled to the United States to begin a lecture tour that was scheduled to last eight months. In September, however, World War II broke out, and in April 1940 Norway was overrun by German soldiers. The occupation would last five years, and by then "Jack" Bjerknes would be the chairman of the new Department of Meteorology at the University of California, Los Angeles. Even as Americans debated in 1940 whether to enter the war, some farsighted military preparations were under way. Bjerknes accepted an invitation by his host country to organize a crash program at a University of California campus to help train meteorologists for U.S. military services. He chose UCLA for the program, one of five university-based emergency military weather training centers launched in the United States.

In the first class at UCLA was a civilian, Jule Charney, a brilliant young mathematical physicist whose reluctant choice of meteorology would lead to striking breakthroughs in the science. Charney credited Bjerknes's special genius with helping him decide to focus on weather science, despite his own "slow and painful" progress understanding the science. Charney recalled in 1975:

> The atmosphere did not fit into the neat and simple categories I had come to expect from mathematics and theoretical physics. But under Bjerknes' tutelage I learned that, though the atmosphere is disconcertingly complex, it is not hopelessly so. To my admiring eyes, Bjerknes was able to produce order where I had seen no order. This, if I were asked to name a single one of his characteristics, is his most outstanding, the ability to extract from an amorphous collection of observations of an exceedingly complex, turbulent fluid always the relevant, distinguishing and characteristic events and processes. To elaborate on his abstractions from nature is to recite much of the history of modern meteorology. His discovery of the front, the frontal wave and the upper wave each in its turn lay the groundwork for a major step forward in classifying and understanding the behavior of the atmosphere.

Charney would go on to formulate the mathematics necessary to finally realize the dream of Vilhelm Bjerknes and Lewis Fry Richardson: precalculating changes in the atmosphere by modern electronic computers. Jack Bjerknes's 1937 paper, published in German in *Meteorologische Zeitschrift,* connecting the upper winds to cyclones at the surface, led directly to Charney's mathematical breakthrough. After the war, observing this fundamental shift in the science—the birth of Numerical Weather Prediction—Jack Bjerknes would choose his own, different direction.

At the age of 60, inspired by Rossby and others, he launched a major new theoretical undertaking that was especially suited to his famous talent for drawing order out of apparent chaos. For 15 years he would conduct an investigation into the long-term, large-scale interactions between the oceans and the atmosphere. He explored the Atlantic's Gulf Stream and then turned his attention to the Pacific, where he achieved one of the most important discoveries of modern climate science.

A major advance would come as he began working with data gathered during the cruises of the International Geophysical Year (IGY) of 1957–1958. This just happened to be the time of an El Niño episode, and the scientist was trying to pull together what was known about the enigma of the warming sea. Up to this period, El Niño had been seen as a local condition—a sea warming confined to the coast of South America, or a seasonal southerly warm current that because it began about Christmas, led Peruvian fishermen to give it the Spanish name for the Christ Child. Measurements from the cruises of the Geophysical Year established that El Niño was a much larger phenomenon.

Another unresolved set of data interested the great theorist. In the early part of the century, the British mathematician Sir Gilbert Walker, while doing research in India on monsoons, made an interesting but improbable statistical connection. Walker found a regular opposite relationship between sea-surface air pressure at one side of the tropical Pacific—at Darwin, Australia—and air pressure near the Central Pacific—at Tahiti. He called this high-low seesaw relationship the Southern Oscillation. Something else caught his attention, although he had no way of explaining it at the time. Walker found a statistical connection between the Southern Oscillation in the atmosphere and far-flung weather conditions, such as warm temperatures in western Canada. Colleagues derided him for such a notion at the time. The idea was not pursued; rather, it was noticed occasionally but quickly dismissed as a statistical correlation that had no physical basis.

Some 50 years after Walker's research, Bjerknes saw the Southern Oscillation through new eyes. In a great intuitive leap, he articulated the true character of El Niño in a seminal article on the subject in 1969. Bjerknes recognized for the first time that the atmosphere's Southern Oscillation, the seesaw of air pressure, and the ocean's El Niño, the eastern Pacific's warm water, were different features of the same oceanwide "coupled" phenomenon. Bjerknes for the first time described the continuous "feedback loop" created by the El Niño/Southern Oscillation (ENSO), as it is often called now by scientists. He called the big pattern of circulation between the atmosphere and the ocean across the Tropics the Walker Circulation. It would be years before science could fill in data supporting his hypothesis, and years more before the link between the El Niño/Southern Oscillation and western Canada's winter's would be explained. But the importance of Bjerknes's insight was understood immediately.

Jack Bjerknes's original insights into the structure of El Niño were published 50 years after his groundbreaking paper that defined middle latitude storms. By then he was emeritus professor at UCLA, a position he held from 1965 until 1975. He died on July 7 of that year at his Santa Monica home from complications after suffering a heart attack.

21

Tor Bergeron

A Gifted Vision

•

VILHELM BJERKNES BECAME the envy of other European scientists for his ability to attract talented students to his small, brave young meteorological enterprise in Bergen, Norway, in the 1920s. While their enthusiasm and loyalty seemed to flow naturally from the charm of his personality and his nurturing ways once they settled in, first persuading them to visit Bergen was part of a very deliberate recruiting campaign. Having painfully watched his father's career in theoretical physics wither in isolation at the turn of the century, Bjerknes keenly appreciated the importance of the social setting of research—what he called the "scientific milieu." This is what had so attracted him to the University of Leipzig before World War I—the exposure of his results to the broader community of European scientists and the ready supply of young assistants interested in working in his Geophysics Institute laboratory. If such a creative milieu were going to take form in Norway, especially in little Bergen, Bjerknes knew from experience that he was going to have to create it himself.

Between the world wars, Bjerknes drew to western Norway a procession of young men who would become some of the most important geophysical scientists of the twentieth century. Among these acolytes was the Swede Carl-Gustaf Rossby, who would become a world-renowned theorist in atmospheric sciences and lead the way to the crowning achievement of meteorology in the twentieth century. And there was the Norwegian Harald Sverdrup, who would become a world famous oceanographer and director of the Scripps Institution of Oceanography in California. But no one came to Bergen more highly recommended, no one was more ardently sought, and no one contributed more critically to the basic formulation and final worldwide acceptance of the new methods of research and new techniques of weather forecasting than a 27-year-old Swedish meteorologist who arrived in Bergen in 1919.

Tor Bergeron had been interested in weather since 1908, when a teacher at his Swedish boarding school assigned him to observe the barometer every day for a month. The daily weather log he began became a habit. The curiosity it provoked became a lifelong investigation. Bergeron was soon monitoring other variables of weather as well as closely observing the sky. He would come to be recognized as an uncommonly gifted observer, especially of the sky's visual aspects. Bergeron learned to discern all manner of atmospheric detail from such features as the transformations of clouds and especially from more subtle changes in visibility. This attention to the sky would become a trademark of Bergeron's methods and powerfully influence the direction of his research long after his days as a principal of the Bergen school of meteorology in the 1920s.

Great vistas had both aesthetic and practical appeal to Tor Bergeron. From 1909—and for most of his life—his family maintained a summerhouse in the Swedish mountains near the Norwegian border. From there he could see across a lake and to the ridges of surrounding mountains, a prospect that offered long studies of leeward cloud formations and air-mass visibilities. Bergeron was a fun-loving, charming man of artistic temperament and occasional Bohemian bent. Sometimes he would arrive so late to his forecasting chores that the radio announcer reading the forecast would have to slow his speech to allow Bergeron to keep pace. He had wide interests and many natural gifts, including music and language. All his life he sang, often accompanied on the piano by his talented wife, Vera, and he participated in academic choirs wherever he lived. He spoke the Scandinavian languages as well as German, French, English, Russian, Spanish, Italian, some Dutch, Czech, and Serbo-Croatian and remembered all of the rules of grammar and usage better than many speakers of the mother tongue. A gifted lecturer, he traveled throughout Europe, educating scientists in the methods coming out of Bergen and elsewhere. It was on such an extended mission to Moscow in 1932 that he met and married Vera Romanovskaya, herself a student of mathematics and physics who often helped Bergeron in his work.

Born in England in 1891, Tor was the only child of Armand Bergeron, a stamp dealer, and Hilda Stawe, a singer. The family returned to their native Sweden when Tor was four years old. Just three years later, in 1898, Armand Bergeron died of rheumatic fever at the age of 36. The widow and child were aided by family members, including relatives of his father who provided the means for Tor's education at the Lundsberg School and the University of Stockholm, where he studied mathematics and physics before graduating in 1916.

While continuing his studies in meteorology after graduation, Bergeron in 1917 and 1918 was developing ideas about changes in visibility and linking them to various atmospheric events. In particular, he was drawing streamlines on his own weather maps and noticing how changes in visibility so obviously followed the passage of lines of convergence associated with storms. This work eventually would become part of his seminal work in air-mass analysis.

Bergeron's talents were observed and encouraged by three important scientists: by a friend of his mother, the meteorologist and explorer Nils Ekholm, who was director of the Swedish weather service; by Johann W. Sandstrom, who had been Vilhelm Bjerknes's first Carnegie assistant in Stockholm; and by Bjerknes's closest friend in Sweden, the Nobel Prize–winning chemist Svante Arrhenius. By the autumn of 1918, Bjerknes had heard about Bergeron from both Sandstrom and Arrhenius and knew that he wanted him in Bergen. When Jacob Bjerknes and Halvor Solberg were sent to Sweden to hunt for young recruits after the first summer of weather forecasting, they had special instructions to find Tor Bergeron.

Many years later, Bergeron would recall the day in November 1918 when he met the Bergen meteorologists. He was sitting in the Meteorological Institute at the Swedish Royal Academy of Sciences, "trying to find out about dust," when Bjerknes and Solberg walked in. "I remember how eagerly Jack and Halvor spoke as they demonstrated with their drawings and their maps how a warm front passed over mountains," he said. "Of course, I had made a lot of cloud observations myself ever since 1908/1909, and as they talked I saw more and more clearly that what I was hearing agreed perfectly well with my own experiences." The Bergen recruiters accomplished their mission, persuading five young men to join them, including Bergeron and Rossby. "So we shall have a living scientific milieu here too," Vilhelm wrote in a letter to Arrhenius. Bergeron went to work for Nils Ekholm at the Swedish weather service in January 1919, but after two months was granted a one-year leave of absence to go study the new meteorology in western Norway.

For young scientists, Bergen was like no other place to work, and Vilhelm Bjerknes was like no other professor. When young recruits arrived at the Bergen train station, there waiting for them was the professor himself, ready to heft their luggage and lead them to their lodgings. "It was something new for us that a famous professor could show such consideration to unknown students," recalled another 1919 Swedish recruit, Erik Bjorkdal, "but that was not enough. Only three days after our arrival he took us on an inspection tour along the coast on board *Armauer Hansen*, so that from the very beginning we should get a living impression not only of Norwegian nature but also of the forecasters' work out in the field." They were soon put to practical work, analyzing weather maps. "Every map was a new problem which had to be solved, and everyone, each according to his ability, contributed towards its analysis," Bjorkdal recalled. "Every time a new secret was tricked from fronts and cyclones there was a great deal of excited interest."

In a letter to a friend, Bergeron described Vilhelm Bjerknes as an "extraordinarily charming man and utterly unassuming for all his genius." Years later, he would remember how Bjerknes nurtured the creative intensity of the young scientists in those early days, when the Bergen forecasting service operated without fixed office hours. "One often sat up the whole night, alone or together with several other young meteorologists, re-analyzed the day's charts, and pondered over or discussed unexpected developments. . . . At times like this V. Bjerknes could break off his mathematical analyses or his writing and appear, eyes flashing in anticipation: 'Are there any new discoveries this evening?'"

Arriving in March, Bergeron quickly made a practical impression on the Bergen Forecasting Service. After the first summer of forecasting for farmers, coastal fishermen asked the Bergen meteorologists to provide a similar winter service to the newly motorized Norwegian fleet. Vilhelm and Jacob were initially reluctant, however, partly because they were unsure of their ability to provide more than large-scale gale warnings. In breaking with the old ways of "pressure meteorology," they had abandoned the use of isobars, which describe patterns of air pressure, in favor of streamlines, which describe the wind field. While this switch had been critical to Jacob's discovery of the new cyclone model, it had another effect on forecasting that the Bergen team did not realize. Streamlines that show convergence were especially useful in forecasting precipitation, which was of interest to farmers, but were not very good for predicting changes in winds, which are closely linked to changes in pressure. Bergeron had just spent two and a half months in Sweden forecasting weather using isobars. He pointed out that including these pressure contours would greatly improve their ability to forecast changes in the direction and intensity of winds, and so provide the kind of detail of special interest to the fishing fleet.

Vilhelm Bjerknes and Halvor Solberg were the stronger theorists among the small Bergen team, but Tor Bergeron was Jacob's equal as a synoptic meteorologist, someone able to discern the character of atmospheric events from the seeming disorder of synchronous observations. In a letter to Sandstrom in October 1919, Vilhelm Bjerknes said Bergeron was "certainly one of the best prognosticians that could be found at the present time, probably the very best of all. He knows how to exploit every single sign either on the chart or in the sky." A month later, Bergeron would employ this skill to develop an important new insight into the centerpiece of the Bergen school of meteorology.

Jacob Bjerknes's 1918 cyclone model was a striking advance over prevailing conceptions of midlatitude storms, which most theorists and weather forecasters continued to view, along the lines of Scottish meteorologist Ralph Abercromby's 1885 model, as a symmetrical circulation around a center of low pressure. Still, the new model was a static depiction of a storm, a snapshot frozen in time. Everyone knew it was moving, of course, and Bjerknes had the idea that he could tell the direction of its travel from the angle of the forward line of convergence in relation to the center of low pressure. For this reason, the forward line of discontinuity, what eventually came to be called the warm front, first was labeled the steering line. But there was no sense of development or evolution going on, no life cycle under way. If the shape of a storm changes over time, the model as Bergeron found it in the spring of 1919 gave no hint of this fact. The two fronts or lines of discontinuity, separated by the wedge of warm air, simply flowed immutably over the landscape.

With his own uncommonly keen gift for bringing order to scattered data, Bergeron soon began to recognize that observations from their dense Norwegian network commonly described a storm that had a structure that was in some way different and more complex than Jacob's model. Young Bjerknes readily conceded the point, but throughout the summer of 1919 he resisted any notion that these differences called for conceptual changes. Although his model was based on the observations of an actual storm, it was only a *model*, after all, and owed to its inherent simplicity much of its value in their campaign to win widespread acceptance of Bergen methods. Tor Bergeron's important breakthrough came in the autumn.

Observations of a storm that swept across Scandinavia on November 18, 1919, crystallized in Bergeron's mind two important dynamic elements that were missing from Jacob's model. First, as Bergeron had suspected, the two fronts of a midlatitude cyclone travel at different rates across the landscape. This meant that Jacob's model of two fronts separated by a broad wedge of warm air represented only a certain phase in a storm's development. Second, as the storm progresses, the faster cold front eventually catches up to and overtakes the leading warm front. The sequence of weather maps for November 18 revealed this circumstance of overrunning taking place in the sky over Norway, where the two fronts were close together near the storm's center of low pressure. Confidently, Bergeron drew a cross-sectional illustration showing the cold air invading and lifting the warm front from the surface. This complex process signals the beginning of the storm's decay, as it deprives the storm's center of its supply of heat and moisture. The Bergen team called this process occlusion, and the compound front of cold air at the surface and warm air overhead became known as an *occluded front.*

After just two years as a weather service meteorologist in Stockholm, Bergeron emigrated to Norway, where he spent the next 13 years, continuing to develop and spread the new meteorology. Better than anyone, he organized the concept of air-mass analysis, classifying air masses according to their origins—equatorial, tropical, polar, and arctic—as well as their type—maritime or continental. More persuasively than anyone, he spread the doctrines of the new meteorology among a skeptical scientific and forecasting community by relating its ideas to current weather conditions. But there was more to Tor Bergeron than Bergen school meteorology.

In February 1922, on his way back to Bergen, he stopped off for a few weeks of recreation at a health resort in the hills north of Oslo, where his observational genius led to another important advance in weather science. The hillside often was enveloped in a layer of stratus cloud. While walking on a road through the fir forest, Bergeron noticed that the fog pattern was not always the same, and that the difference depended on the temperature. When temperatures were well below freezing, the fog did not descend all the way to the roadbed, as it did when temperatures were warmer. "My tentative explanation came immediately," Bergeron recalled. "At –10°C [14°F] the rime-clad branches of the firs along the windward side of the road by diffusion transport 'filtered away' so much of the water vapor in the air . . . that the fog droplets partly evaporated." Bergeron realized that the ice was taking up water vapor, which at warmer temperatures remained in the air. In 1928, in his doctoral thesis, written in German, he mentioned this idea about the role of ice crystals in clouds. By 1933, at an international conference in Lisbon, the idea was fully developed in his paper, "On the Physics of Cloud and Precipitation." His famous ice nucleus theory was to help explain one of the most enduring mysteries of weather.

What causes rain to fall? Scientists could explain the formation of clouds by condensation long before they had a persuasive physical explanation for precipitation. Bergeron was first to show how minuscule water droplets amass themselves into drops large enough to fall as rain. In the cold heights of clouds, he explained in 1933, ice nuclei take up molecules of water vapor from supercooled water droplets. When the ice crystals grow large enough, they begin falling as snowflakes. As they descend into layers of warmer air, the flakes melt and continue to fall as raindrops. This *Bergeron process* explains most precipitation in the middle and higher latitudes and is the scientific basis of ever-hopeful efforts to produce rainfall by artificial means.

Meteorology textbooks are full of concepts to which Bergeron made contributions. Even the daily weather map carries his mark. In 1924, when the time came to begin printing weather maps for widespread distribution, it was Tor Bergeron's idea to abandon the colors blue and red and to adopt symbols that would allow cold fronts to be distinguished from warm fronts in black ink.

22

Carl-Gustaf Rossby

Conquering the Weather Bureau

•

MORE THAN ANYONE, it was Carl-Gustaf Rossby who finally dragged the U.S. Weather Bureau into the twentieth century. It was he who created the first American university programs in meteorology and set the nation on a course of scientific excellence. It was Rossby, more than anyone, who organized the emergency training of thousands of military meteorologists in time to contribute to the Allied victory of World War II. And it was Rossby whose own celebrated research opened the way to extended forecasting and helped achieve the century's great dream of numerically calculating the weather. Arriving in Washington, D.C., early in 1926, unheralded and all but unwelcome, the father of modern American meteorology was not an American but a Swede.

Sponsorship came privately from the American-Scandinavian Foundation, which granted the young meteorologist a fellowship to travel to the United States and to work at the Weather Bureau to study the application of the polar front theory to American weather. He was 27 years old and physically unimposing—he was short and had a round face—although he was vigorous and all of his life an uncommonly hard worker. In personality and intellect, Carl Rossby was everything that chief Charles F. Marvin and the other old men administering the Weather Bureau were not. He was trained in mathematics and physics and scientifically astute. He was brilliant, charming, persuasive, and relentlessly enthusiastic about spreading the new science of the Bergen school of meteorology across the United States. At the central office of the Weather Bureau, where these characteristics were not in great demand, the new visitor from Scandinavia was assigned a desk that was located in a far corner of the agency's library. Even this remote station would be withdrawn before long, and he would become persona non grata—an official posture, no doubt, which inspired false hopes by the U.S. weather bureaucrats that they had seen the last of Carl Rossby.

Born in 1898 in Stockholm, Carl was the first of five children of Arvid and Alma Charlotta Rossby. His father, a quiet and jovial man, was a construction engineer of modest means. His mother was the daughter of a wealthy pharmacist in Visby, which is located on Gotland, a Baltic Sea island famous for its orchids. Visits to Gotland, shared later with American friends, inspired an early interest in botany and a lifelong interest in orchids. To spend time with Rossby was to live in the wake of an electrifying personality. He was bursting with a delighted, infectious interest in many things, and he liked nothing better than sharing. But there was a private side to this thoroughly social creature. Even close friends did not know of his interest in music or his early talent at the piano. And some secrets were darker. Rossby did not share with friends the experience of his affliction with rheumatic fever as a boy or the medical knowledge that the disease had seriously damaged his heart.

Rossby obtained a bachelor's degree at the University of Stockholm in 1918, specializing in mathematics, astronomy, and mechanics, accomplishing in less than a year a course of studies that was supposed to require three. Professors noticed this young, precocious, inquisitive, restless student. In the winter of 1918–1919, when Vilhelm Bjerknes revisited the university while scouting Sweden for young talent to fill his needs at his new Geophysics Institute and weather forecasting service in western Norway, Rossby's name was high on the list. Eager for adventure, but without a stitch of education in meteorology, Carl Rossby arrived in Bergen during the exciting summer of 1919, just as the polar front theory was taking shape.

"This boy of 20 had an amazing persuasive and organizing faculty," Tor Bergeron would recall of Rossby's first days in Bergen. In a Rossby memorial volume, *The Atmosphere and the Sea in Motion,* Bergeron wrote "His far-reaching ideas and high-flying plans often took our breath away. Soon he was also able to make practical suggestions of value for the experimental weather service that was connected with the Geophysical Institute at Bergen." Still, for all the intellectual ferment of those days, Rossby may have felt somewhat out of place in Bergen. The engaging informality of those days, so different from the strict formalism of Swedish academia, was a style that Rossby would emulate everywhere he went, but this may have been the one occasion in his life when he did not feel intellectually in command of his subject. He was not especially well equipped for, or interested in, the routine technical mapping of data that was central to practical daily forecasting. At the same time, while he naturally gravitated toward theory, he lacked the knowledge of higher-level mathematics and physics that might have allowed him to contribute some dynamical understanding to the work that was under way that summer. As Bergeron observed, "Rossby had not yet, at Bergen, found the time and place, and the branch of meteorology, where his special capacity as a scientist and organizer could prosper." Certainly he had fallen in love with the science of meteorology, and he was thoroughly indoctrinated in the new ways of air-mass analysis. After a year, however, he was ready to leave Bergen, to begin to find his own way, but first he had to secure more advanced scientific skills.

Following in Vilhelm Bjerknes's footsteps, Rossby went to the University of Leipzig, where the young Swede first encountered valuable data from direct soundings of the upper air. He spent part of 1921 near Berlin at the aeronautical observatory in Lindenberg, a center for kite and balloon observations. Later in 1921, he returned to the University of Stockholm to study more mathematical physics. To earn money for his education, he took a year off to work as a meteorologist with the Swedish weather service. During the summers, he served as the meteorologist aboard oceanographic expeditions. In 1923, he was on board the small Norwegian vessel SS *Conrad Holmboe* when it was caught in pack ice off the shores of eastern Greenland. For two months, the vessel was helpless and in danger of sinking. The cruise almost ended in disaster before help arrived. In 1925, Rossby obtained the equivalent of a master's degree in mathematical physics from the University of Stockholm, and his formal education came to an end. Rossby's intellectual restlessness got the better of him. The scientist who for 25 years would be the single most influential figure in meteorology in the United States never did obtain a doctorate.

After arriving in Washington, D.C., in 1926, Rossby soon found in meteorologist Richard H. Weightman one of the few people at the U.S. Weather Bureau interested in the new ideas from the Bergen school. The two collaborated on a paper that appeared in the December issue of the bureau's *Monthly Weather Review:* "Application of the Polar-Front Theory to a Series of American Weather Maps." A classic of its kind, the article concluded that their research "has furnished conclusive evidence that the polar front theory can be applied with great advantage to even rather complicated weather maps in the United States and that it enables us to explain phenomena which without a knowledge of the dynamics of the situation would hardly be understood."

This paper had no more impact on the Weather Bureau than any of the science appearing in its monthly publication. With missionary zeal, Rossby was preaching to men who were 40 years his senior, forecasters that meteorologist Horace R. Byers later would describe as "half-educated practitioners who had risen through the ranks because of some practical knowledge and ability to outguess the weather." In charge was Charles Marvin, an old bureaucrat who trained as an instrumentation engineer, and whose career in the agency dated back to the 1880s, when the weather service was part of the U.S. Army Signal Corps. In this deeply entrenched bureaucracy, neither Chief Marvin nor anyone else in authority saw reason to change the way the Weather Bureau had been doing things for decades. These men, observed Byers, "had no interest in Rossby's scientific brilliance but rather found the young Swede, with his schemes for revitalizing meteorology in the United States, a great nuisance."

Outside of the Weather Bureau, however, the burgeoning field of aviation was generating urgent new practical interest in the state of the science of meteorology in the late 1920s, and the advantages of the new forecasting methods of the polar front theory seemed obvious. Among the first to discover Rossby hidden away in the library was a young U.S. Navy lieutenant, Francis W. Reichelderfer, who was in charge of Naval Aerology and made daily data-gathering visits to the Weather Bureau to help prepare his own forecasts. Reichelderfer was delighted to find someone so familiar with Bergen school methods. The two young men, close in age, spent hours together after work in critiques of the day's weather maps. And Reichelderfer introduced Rossby to his friend Harry F. Guggenheim, president of the Daniel Guggenheim Fund for the Promotion of Aeronautics. Just as Rossby's American-Scandinavian Foundation grant was running out at the Weather Bureau, Harry Guggenheim, himself a World War I pilot, arranged to extend his financial support. It was Reichelderfer and Guggenheim who recognized Rossby's promise and set the stage for remaking meteorology in the United States.

Early in 1927, with Guggenheim's support, Rossby helped prepare weather forecasts when Richard E. Byrd crossed the Atlantic in the three-motored *America* with a crew of three a few weeks after Charles A. Lindbergh's famous solo flight to Paris in the *Spirit of St. Louis*. In July 1927, while still at the Weather Bureau, Rossby was named the full-time, paid chairman of the new Guggenheim-funded Interdepartmental Committee on Aeronautical Meteorology. From this position, he arranged to send a young and promising Weather Bureau employee, Hurd C. Willett, to a year of study in Bergen.

As Rossby's reputation began rising outside of the agency, what Byers described as "a series of minor incidents" in connection with his Guggenheim work led to increasing conflict with Marvin and other Weather Bureau administrators. The final incident came after Rossby received a call from Charles Lindbergh. Sponsored by Guggenheim, the new American hero had just completed a 48-state tour in the *Spirit of St. Louis*, landing in 100 towns across the country and preaching his vision of the future of aviation to millions of Americans. Now Lindbergh had been persuaded to attempt a 27-hour nonstop flight from Washington, D.C., to Mexico City. The famous pilot, whose low opinion of Weather Bureau forecasts had been widely publicized, asked Rossby to forecast the weather between Washington and Mexico City for December 13–14. After successfully completing the flight, Lindbergh called Rossby's the best weather forecast he had ever received. So incensed was Marvin by this unauthorized forecast that Rossby was told he was no longer welcome at the Weather Bureau.

But Lindbergh's flights and his unprecedented Guggenheim tour had awakened public interest in traveling by airplane, and now Harry Guggenheim had bigger plans for Rossby. To show the public that flying in airplanes was not just for daredevils anymore, the Guggenheim fund was financing the operation of a model airway service in California, to run between San Francisco and Los Angeles. Run by Western Air Express along its existing airmail routes, the model San Francisco–Los Angeles passenger service needed to prove to an uncertain American public that air travel was safe and reliable. This would require the best weather forecasting service possible. Rossby was asked to develop a model weather service to support this first regular airline passenger service in the United States, which would open for business in the spring of 1928.

It was Rossby's first major organizing effort, and he plunged in with characteristic zeal and flair for leadership. He walked into the San Francisco Weather Bureau office soon after the local official in charge, Major Edward H. Bowie, had received a letter from Marvin warning that Rossby was not to be accorded any special services or access to the offices of the agency. A wiser and more forward-thinking man, Bowie disregarded the letter and welcomed Rossby to California. With Western Air Express pilots, Rossby flew up and down the Central Valley, landing outside small towns with much fanfare, dining with mayors and local businessmen, and gaining their help in securing the most reliable local observers, who would report to the weather collection center at the Oakland Airport. Unlike the

Weather Bureau's simple arrangement, Rossby created a dense and broadly arrayed observation network, recognizing that bad flying weather was likely to sidle in from surrounding mountains rather than along the north-south axis of the route. Joining the 29-year-old Rossby in this project was Horace Byers, a young University of California student, who would become a protégé and a valuable managerial assistant. The weather service proved successful beyond all expectations, and the model airway marked the beginning of the age of passenger air travel in the United States. Not a single airplane accident was caused by weather during the experimental year. The model weather forecasting service won rave reviews from military as well as civilian pilots in California. Recognizing the inevitable, the U.S. Weather Bureau agreed to officially take over the service on July 1, 1929, and eventually extended it across the nation.

Carl Rossby, meanwhile, had already moved on to an enterprise that would establish his credentials, at the age of 30, as a leading figure in American meteorology. Guggenheim had funded the creation of a department of aeronautical engineering at the Massachusetts Institute of Technology. In 1928, Rossby was named associate professor of meteorology at MIT, with responsibility for organizing and implementing the first complete academic program in weather science at an American university. Over Marvin's objections at the Weather Bureau, Rossby enlisted for his faculty young Hurd Willett, who was freshly returned from a year's study in Bergen. The first year of instruction drew Reichelderfer's naval officers, and the second year began attracting civilian students, including Horace Byers from California, who was the first civilian to receive a doctorate in the program. Here Rossby met Harriet Alexander, of Boston, and the couple were married at the beginning of his second year at MIT.

"At MIT Rossby exhibited that leadership for which he was famous," Byers would later write. "Those who studied under him practically worshipped him. They were participating in his great crusade—to bring modern meteorology to America where the science had been existing in a stifling atmosphere for many years." Byers described the experience as exhilarating: "His lectures were carefully prepared and given with enthusiasm and his informal discussions . . . were nothing less than an inspiration."

Rossby would remain at MIT for 11 years, longer than anywhere else, and during that time he would singularly transform American meteorology from a wasteland into a leading international center of the science. Like Vilhelm Bjerknes's "living scientific milieu" in Bergen, Rossby's department of meteorology was noted for its informality, the freewheeling intensity of its scientific discussions, the stellar international talent it attracted as visiting faculty, and the close link it maintained between meteorological theory and the practice of weather forecasting. With a changing of the guard at the Weather Bureau, MIT's meteorology department would educate key new personnel. Like Bergen, the daily map discussion was a centerpiece of Rossby's academic program at MIT. Always the discussion led from the theoretical to the practical. "The proof of the pudding," he would say, "lies in eating it."

Meteorological history of another kind was made at MIT. In the midst of far-flung organizing work, which engaged him now everywhere he went, Carl Rossby presented results from his own research that established his reputation internationally as a theorist of atmospheric sciences second to none. In 1937, Jacob Bjerknes had published a paper that identified the formation of cyclones with hemisphere-scale upper-level waves. Extending this thought in 1939 and 1940, Rossby went much further, reaching what meteorologist George P. Cressman would describe as "a stunning conclusion." The upper-air waves extend vertically through the lower atmosphere, Rossby said, and the movement of the waves—which cause the movement of warm and cold air masses that produce local weather—is powered by the distribution of the forces derived from Earth's rotation. Rossby's famous equation accounts for this rotational energy, known as *vorticity*, because air currents as well as ocean currents flow from one latitude to another. In August 1939, in a paper presented in Toronto titled "Planetary Flow Patterns in the Atmosphere," Rossby concluded that "the factors determining the stationary or progressive character of the motion are to be found in the vorticity distribution and that the displacement of the pressure field is a secondary effect." In a 1996 interview, in the *World Meteorological Organization Bulletin,* Cressman called this contribution by Rossby "the key to the coming revolution in dynamic meteorology and in forecasting. The question that was to occupy us in the future was at hand; namely, to understand the full significance of Rossby's conclusion."

The effect of this rotational force creates a pattern of four to six large-scale "planetary waves" around the hemisphere that define the pattern of high-level westerly winds that encircle the pole and mark the boundary between polar air and warmer tropical air. These semipermanent waves, as well as their ridges and troughs, which would become such familiar features of modern weather maps, came to be known as Rossby waves, and are found in ocean basins as well as the atmosphere. Employing his famous talent for ruthlessly simplifying complex problems, Rossby went on to devise a mathematical formula that allows theorists and forecasters to calculate the movement of these waves—and the weather they carry—around the hemisphere. Extended five-day forecasts were first made possible by the use of this Rossby equation, which, according to science historian Gisela Kutzbach, "became perhaps the most celebrated analytic solution of a dynamic equation in meteorological literature."

In 1939, Rossby was naturalized as a U.S. citizen. Also that year, Reichelderfer was appointed chief of the U.S. Weather Bureau, with a mission and a mandate to modernize the nation's weather service. Five years earlier, a special scientific panel appointed by President Franklin D. Roosevelt had urged the agency to adopt the Norwegian air-mass analysis, but even this high-profile admonishment had not changed daily forecasting practices. Chief Marvin finally had resigned, and since 1935 Horace Byers and a few other MIT graduates had been at the agency, trying to initiate Bergen school forecasting methods. Meteorologist Jerome Namais recalled the continuing resistance to Byers's efforts. "His group was placed in a corner of the Weather Bureau—a safe distance from the forecasters," Namais wrote. "The forecasters were doing 'the real thing' and could not be contaminated by the young upstarts analyzing strangely in the other room." Real change did not come until Reichelderfer, the first outsider to head the agency in nearly 50 years, was put in charge. At Reichelderfer's urging in 1939, Rossby reluctantly agreed to leave MIT and temporarily accept the position of assistant chief for research and development of the U.S. Weather Bureau. Twenty years after Jacob Bjerknes's seminal paper on the subject, the irresistible forces of Reichelderfer's determination and Rossby's persuasive skills finally overcame the institutional resistance of the U.S. Weather Bureau to new scientific methods. In Washington, Rossby helped put a major reeducation program in place while Byers went off to Chicago to lead one of the regional training centers. But Rossby was not interested in government work, and in 1941 he left to become chairman of a new meteorology program at the University of Chicago, where Byers became an associate professor and Rossby's strong administrative hand—an arrangement which lasted for years.

More urgent matters would intervene, however. War had begun in Europe. The Nazis had invaded Norway in April 1940, stranding Jacob Bjerknes in the United States, and the likelihood of U.S. involvement in a second world war was growing each day. Carl Rossby was among the first to realize that this war would involve enormous airpower as well as weapons of mass destruction that would require battlefield weather forecasting. In 1940, even as public political dialogue focused on staying out of war, the U.S. government organized a massive training program for military meteorologists. It was Rossby's powers of persuasion that led military leaders, against their inclination, to employ civilian universities in the training of wartime weather forecasters. More than 7,000 servicemen underwent emergency training in meteorology at MIT, New York University, the University of Chicago, UCLA (where Bjerknes had established a department), and Caltech, as well as key military centers. The war years found Rossby traveling to every part of the world, helping local British and American military forecasters deal with special problems they encountered in unfamiliar skies. When military meteorologists realized that they knew too little about the Tropics to be effective in that theater, Rossby helped organize the University of Chicago's Institute of Tropical Meteorology in Puerto Rico.

The postwar years brought a flowering of scientific advances under Rossby at the University of Chicago, representing a leap of progress that probably was equal in importance to the Bergen school breakthroughs of the 1920s. Rossby's long-wave formula had crossed the divide from synoptic analysis of regional features to the kind of planetary dynamics that Vilhelm Bjerknes described when he dreamed of making meteorology an exact science in 1904. In the late 1940s in Chicago, one protégé after another plowed the ground that would lead to Numerical Weather Prediction by electronic computer, the triumph of weather science in the twentieth century. Byers was at his side, tending to administration, a role that cost him a certain amount of scientific recognition in his own right. The meteorologist H. Robert Simpson, an alumnus of that time, later would describe Byers as "the balance wheel in the administration of one of the great test meteorology programs the world has ever known; a spirited, if at times unruly, department energized in its early days by the creative genius of Carl Rossby." At the University of Chicago, Rossby was at his theoretical and inspirational best, building another remarkable academic research program around dynamic concepts. In an echo of Bergen, the program became known as the Chicago school, or the Rossby school, of meteorology.

What meteorologists think of as the famous "dishpan" experiments began in Chicago when Rossby encouraged graduate student Dave Fultz to "explore turbulent mixing using the hemispheric shell." Large round pans set on turntables were filled with water, chilled at their centers, and heated at their edges; their carefully controlled rotation proved remarkably effective at simulating the general circulation of the atmosphere. In Chicago after the war, the visiting Finnish meteorologist Erik Palmen, Rossby, and others defined the *jetstream*, a meandering current of high-speed winds embedded in the high-level westerly winds that had surprised

American bomber pilots crossing the Pacific during the war. Rossby and his collaborators—students and faculty, Americans and numerous international visitors—laid the theoretical framework for the numerical simulation of a simplified atmosphere that proved to be critical to the success of Numerical Weather Prediction by electronic computer in the 1950s. It was Rossby and friends who encouraged the great mathematician John von Neumann to take on the problem of Numerical Weather Prediction as one of the first major tasks for his new electronic computer. In 1947, however, just as that project was beginning to come to fruition at the Institute for Advanced Study in Princeton, New Jersey, Rossby surprised and disappointed many of his American colleagues and protégés by accepting the urgings of his native nation and returning to Sweden.

It was time to go. Nobody ever understood quite why. Carl Rossby was at the top of his powers scientifically and intellectually at the time. Perhaps he thought the return to Sweden would allow him to slow his hectic pace. He always felt that moving on to new domains and fresh challenges was vital to his personal development. Again during this time he surprised colleagues by taking up an entirely new set of scientific questions involving atmospheric chemistry and its effects on climate and weather. In this regard, Rossby was among the first meteorologists to recognize the potential harm of the continued industrial pollution of the atmosphere. Appearing on the cover of *Time* magazine in 1956, Rossby expressed misgivings about the buildup of various airborne substances, and noted that the atmosphere is "man's milieu" and that his well-being is linked so intimately to it. "Tampering can be dangerous," he warned. "Nature can be vengeful."

On August 19, 1957, Carl-Gustaf Rossby died suddenly in his office of a heart attack. He was 58 years old and at the top of his international fame and influence. The loss was personally staggering to the hundreds of meteorologists he had inspired around the world—none more so than to young Jule Charney, the brilliant theorist who was implementing Numerical Weather Prediction at the Institute for Advanced Study. Returning from Stockholm, he met a fellow meteorologist at a Paris airport and told him the awful news. "Without Rossby," Charney said, "my world has become dim."

23

Sverre Petterssen

Forecasting for D-Day

•

TWO MILLION MEN were waiting on six. In a state of what their commander called "suspended animation," combat troops intent on liberating Europe from the armies of Adolph Hitler were waiting for a group of weathermen to decide if conditions would permit the largest military invasion in history to go forward on June 5, 1944—or not. Airmen were waiting to fly 11,000 aircraft carrying bombs and paratroopers. Sailors were waiting in 600 warships that would escort American, Canadian, and British soldiers who were waiting in more than 4,000 ships and landing craft to cross the Channel from southern England to France and land on the beaches at Normandy.

They were waiting on Geoffrey Wolfe and Lawrence Hogben of the British Admiralty, who were two floors underground in downtown London; on Sverre Petterssen and C. K. M. Douglas of the British Meteorological Office at Dunstable northwest of London; and on Ben Holzman and Irving P. Krick of the U.S. Strategic Air Forces in suburban Teddington. That three separate weather centrals were involved was a measure of the enormity and the international character of the undertaking, as well as the critical role that weather conditions would play. A fourth unit comprised of Chief Meteorological Officer James Martin Stagg and his deputy Don Yates coordinated the weathermen and actually briefed General Dwight D. Eisenhower, the Allied commander, and the rest of the military brass.

The weather had been "invasion-perfect" throughout most of May, while a shortage of landing craft was being filled and tactical air strikes were hammering German airfields in France. And the first few days of June were fine and bright when all was in readiness and the waiting was becoming nearly unbearable. Half a world away, however, high above the surface were the first signs that a profoundly indifferent atmosphere could be conspiring against the commander and his armies of liberation.

"We wanted to cross the Channel with our convoys at night so that darkness would conceal the strength and direction of our several attacks," Eisenhower wrote later in *Crusade in Europe*. "We wanted a moon for our airborne assaults. We needed approximately forty minutes of daylight preceding the ground assault to complete our bombing and preparatory bombardment. We had to attack on a relatively low tide because of beach obstacles which had to be removed while uncovered. These principal factors dictated the general period; but the selection of the actual day would depend upon weather forecasts." Everyone had their own ideas of perfect weather. Bomber pilots wanted clear skies, and paratroopers wanted the protection of cloud cover. Ground forces wanted onshore winds, and naval forces wanted the small waves of offshore winds. But there were strict operational limits, and a storm could spell disaster. In high winds and large sea swells, this giant amphibious invasion could founder and fail. On the basis of the only entirely predictable elements—phases of the Moon and low morning tides—Eisenhower had little to chose from: the three-day period beginning June 5. Without good weather on one of those days, he feared, "consequences would ensue that were almost terrifying to contemplate."

First to sound the alarm was Petterssen at Dunstable. This renowned Norwegian scientist had formed a special branch at the British Met Office that was focusing on relatively new information about the upper air now coming in from *radiosonde* balloons and aircraft reconnaissance. On May 28, he didn't like what he saw. During a conference call late that Sunday night, when Stagg asked for the five-day outlook, Petterssen warned that the upper-air situation was becoming unstable across the Atlantic. The high westerly winds were gaining strength. By Friday, June 2, deterioration could set in. At the Admiralty, the naval forecaster Wolfe shared these concerns, whereas Krick, the U.S. forecaster, saw nothing but continued quiet weather throughout the five days.

Petterssen and Krick argued their cases forcefully. Under most circumstances, Petterssen would not often hazard a forecast of such long term, but he thought he saw an exceptionally strong and troublesome pattern taking shape. As he recalled in his autobiography, *Weathering the Storm,* Petterssen couldn't yet say which day would be better or worse, but he recognized the risk the new pattern could pose to what he regarded as "the do-or-die question of the navy: Can we, or can we not, land the force in fighting shape on the coast of France?" The American Krick, on the other hand, was always extravagantly confident about his skills as a long-range forecaster. Krick had studied weather charts and data going back 50 years and was certain that the Channel would be protected by the Azores High, a pressure pattern that commonly dominates the eastern Atlantic during summer. Back and forth they argued, and Stagg, who was more manager than meteorologist, grew weary of the battle. Stagg told his version of events in his book, *Forecast for Overlord.* He wrote of Petterssen and Krick that "each was confident in his own diagnosis and prognosis, each had an aptitude for dogmatic assertions, relaxation from which became harder as the discussion progressed."

The realm of forecasting weather for the next five days and beyond—especially English weather—was a scientific netherworld more congenial to doubt than certainty. At Dunstable, the veteran British forecaster Douglas, who knew more than anyone about English weather, wanted nothing to do with forecasts beyond 48 hours. The meteorologists were routinely providing five-day forecasts only because the military demanded them.

More surprising than the differences in the long-range forecasts was the unusual and troublesome circumstance that they did not naturally resolve themselves in the next few days. Paradoxically, as the range of the forecasts became shorter, as more and more data filled the void, conflicts instead became more sharply defined. Behind the scenes, during long telephone conferences over scrambled signals, critical differences among the weather forecasters, especially between Petterssen and Krick, were fought out in an atmosphere of increasing tension and occasional hostility. Meteorological conditions over the North Atlantic were complex and difficult to forecast accurately and certainly unusual for early June of 1944, but something else was at work. Two very different approaches to the problem of weather forecasting were doing battle in the hands of two very different individuals.

Sverre Petterssen was a 46-year-old son of Nordland, the Norway above the Arctic Circle, where his grandfathers had plied the fish-rich waters in shallow-draft open boats designed by Vikings. He knew the North Atlantic like few others, and he understood meteorology, weather analysis, and forecasting, in particular, better than just about anyone. A student of the Bergen school, he had succeeded Jacob Bjerknes as head of the regional forecasting service for western Norway. During those years he had tutored Francis W. Reichelderfer, the U.S. Navy officer who in 1938 had become chief of the Weather Bureau, in the Bergen techniques of air-mass analysis. Petterssen had been a professor of meteorology at the

Massachusetts Institute of Technology, succeeding Carl-Gustaf Rossby, and was the author of the first textbook on the subject of weather forecasting. He was an internationally respected European scientist and a proud man, especially proud of his scientific heritage.

The Nazis' invasion of Norway in April 1940 had stranded Petterssen while he was at MIT on leave from the Norwegian meteorological service. He had quickly put himself at the disposal of the Allies. On both sides of the Atlantic, his services as a researcher and a combat weather forecaster were in much demand. At the British Forecasting Center in Dunstable, Petterssen's Upper-Air Branch in 1944 was doing groundbreaking work in the use of newly acquired upper-air data for weather forecasting, which would prove to be of long-term value to military and civil aviation.

Irving P. Krick was 37 years old in the spring of 1944. He was a brilliant American salesman, and weather forecasting was his product line—although, like many a great salesman, his number one product was himself. Born in San Francisco, he received a bachelor's degree in physics from the University of California, Berkeley, although science was not his first choice. An accomplished pianist, he embarked on a career in music, but found it financially unrewarding. He held jobs briefly at a radio station and a stock brokerage until the Great Crash of 1929. Weather became his focus after his brother-in-law, Horace Byers, Rossby's MIT graduate student, told him it paid better than music. For some reason, he chose the California Institute of Technology in Pasadena, which offered some mathematics but only a few meteorology courses as part of its aeronautics department. He studied under the seismologist Beno Gutenberg, who taught a course in atmospheric structure, and Theodor von Kármán, the aeronautical engineer. Both quickly groomed Krick as their first doctorate in meteorology, which he received in 1934. Krick made a name for himself with a study that argued, after the fact, that the 1933 crash of the U.S. Navy airship *Akron,* which killed 74 airmen, had been the result of a forecasting mistake by the Weather Bureau. Here he befriended General Henry H. "Hap" Arnold, chief of the Army Air Forces. Arnold was a colonel at the time and was stationed at nearby March Air Force Base. Krick sold Arnold on his brilliance as a forecaster. Within just three years, Krick was head of a new Caltech meteorology department that was uniquely focused on meeting the profitable forecasting needs of industry—not just aviation, but the Hollywood movie studios and anybody else Krick could sell on his ideas. In 1935, Caltech invited Petterssen to California for four months of lectures to help attract students and add some theoretical credentials, although meteorological theory was never a big part of the program. Caltech meteorology was about Irving Krick and his thriving forecasting business. Krick was smart, persuasive, optimistic, confident, and singularly ambitious. About one long-range forecast he could be especially certain: he was going to make a lot of money.

During the war, Hap Arnold's Army Air Forces became an enthusiastic employer of Krick's forecasting methods. While others were trying to understand why weather changes, Krick was trying to outsmart it. Whatever their causes, he was on the lookout for patterns, cycles, and types. The key to weather's future was its past, Krick believed: find an old weather map that most resembles current conditions, and succeeding maps most likely reveal how the weather will change in the days ahead. Critics called this analog method the "canned memory" approach to forecasting, or worse. When George P. Cressman, fresh from the meteorology programs at New York University and the University of Chicago, arrived at Air Force Weather Central in Great Falls, Montana, he was told he would be teaching the Krick method. Cressman recalled the occasion in a 1992 interview conducted under the auspices of the American Meteorological Society. "Well, you can't be serious," he remembered saying, "that's a bunch of crap!" Cressman, who would go on to become director of the National Weather Service, quickly found himself transferred. Rossby, who was at the University of Chicago and served as an adviser to Secretary of War Henry Stimson, and Chief Reichelderfer at the Weather Bureau were appalled at Krick's appointment to the U.S. forecasting team in the spring of 1944. Rossby in particular detested his methods and his brash salesmanship, and warned that he had no standing as a scientist. But as Krick had proven more than once to anyone in his way, in the U.S. Air Weather Service his standing with Hap Arnold was all he needed.

Now the D-day clock was ticking down. The weather was becoming more complicated. The behind-the-scenes conflicts among the weathermen were becoming more critical. On Sunday, June 4, the commanders in chief of the largest invasion force in history were waiting for James Martin Stagg to deliver to General Eisenhower a single, accurate weather forecast for June 5. The tall, dour Scotsman, a civilian administrator given to self-doubt and an apologetic manner, was out of his depth scientifically and temperamentally ill-suited to the occasion. He was fortunate to have Colonel Donald N. Yates of the U.S. Army Air Forces as his deputy. A West Pointer and an outgoing, confidant, and strong-willed military man, Don Yates had learned meteorology from Irv Krick. But Don Yates already had confronted Hap Arnold on this touchy subject, and he knew how to handle Krick. "I realize you think Dr. Krick is God as far as meteorology is concerned," Yates remembered telling his commander. "As far as I am concerned, he is part of the military organization and is going to have to act like that, or I don't want anything to do with it." Arnold had replied, "Whatever you say is all right for this round." In the final hours before June 5, it was probably the strength of Don Yates—and the science of Sverre Petterssen—that saved the D-day invasion from what could have been a terrible disaster.

"If ever in the history of weather forecasting there was an occasion for unanimity of view and confidence in the outcome, this was it," Stagg wrote in *Forecast for Overlord*. "Instead, here was a deep cleavage and uncertainty." It was 3:00 A.M. on Sunday, June 4. The weather was clear and winds were calm over southern England, the Channel, and the beaches of Normandy. But the chart of the North Atlantic looked like the middle of winter. Eisenhower and the commanders were expecting Stagg at 4:15 A.M. for a weather briefing before making the final decisions about June 5.

Stagg had warned Eisenhower on Friday night that "the whole situation from the British Isles to Newfoundland has been transformed in recent days and is now potentially full of menace." He had said nothing of the deadlock among the weathermen, and Yates had not intervened to present Krick's rosy forecast of relatively clear skies and stable conditions for D-day. Leaving the room Friday night, the chief meteorological officer, his stress punctuated by hollows under his eyes, had missed the remark by Admiral of the Fleet Sir George E. Creasy: "There goes six foot two of Stagg and six feet one of gloom."

Losing trust and patience in his own forecasters, Yates on Saturday had ordered Holzman and Krick to get in touch with Dunstable and find a way to bring their forecasts more in line with one another. Although Krick resisted, the pressure from Yates nevertheless resulted Saturday night in the most nearly unanimous forecast in several days. Briefing Eisenhower, Stagg described the whole weather setup as very disturbed and complex. "We cannot have much confidence in what will happen and how it will happen from day to day," he said. "Even for tomorrow the details are not clear. But we do know now that the extension of the Azores anticyclone towards our southwest shores, which some of us thought might protect the channel from the worst effects of the Atlantic depressions, is now rapidly giving way." A series of three storms was bearing down on the region. "They are all vigorous and one or more of them may have further intensified by the time they reach the British Isles," Stagg noted.

Along the lines of the presentations by Dunstable and the Admiralty—and in stark contradiction to the outlook suggested by Krick—Stagg on Saturday night had issued a negative forecast for June 5. A cold front would pass through late Sunday or Monday. The day would be overcast and stormy. The winds in the Channel would be too strong for troops to land successfully on the beaches of Normandy, and the clouds would be too thick for bombers to see their targets. Yates said nothing in defense of Krick's contrary views.

Eisenhower gave himself seven more hours.

At the 3:00 A.M. telephone conference, the Admiralty forecaster and Petterssen at Dunstable found nothing in the latest data to change their forecasts for June 5. Krick and Holzman, however, continued to argue against the pessimistic forecasts. They had studied every June weather map for the past 50 years, they said, and every analogue suggested that a ridge of high pressure would protect the Channel from the brunt of the oncoming storms. The other weather forecasters were not persuaded. Neither was Stagg and neither was their boss, Yates. After the 4:30 A.M. weather briefing from Stagg, Eisenhower postponed the D-day invasion for at least 24 hours.

During this briefing, General Carl A. "Tooey" Spaatz, commander of the U.S. Strategic Air Forces in Europe, was especially struck by the discrepancy between Stagg's forecast for Ike and the outlook that Spaatz had received independently from Holzman and Krick. He would mention this sometime later to Yates.

"Yes, and we're very fortunate that that's the way it went," Yates answered, referring to Eisenhower's decision to postpone D-day, "Individuals can make mistakes."

"Yes, they sure can," Spaatz replied.

Krick and Petterssen continued to argue, but the weather centrals managed somehow later Sunday and Monday to cobble together an agreement that behind the storm front would be a propitious period of relatively calm winds and seas and clearing skies before another storm front blew through. Conditions on Tuesday, June 6, would not be ideal, but they would be good enough and hold long enough for the great invasion force to get ashore. And so they were, and so it did.

In the midst of calm and clear weather over southern England and the Channel, the Allied supreme commander and the commanders in chief had trusted their forecasters that bad weather was coming and had postponed the D-day invasion. Thick, low clouds had swept in, and high seas and gale-force winds had lashed the region. In the midst of the storminess, the military leaders again had put their trust in the weather forecasters' expectations that better weather was on the way and had decided to launch the assault that began the liberation of Europe. Petterssen would regard it as meteorology's finest hour.

In the afterglow of victory, American journalists would welcome an American hero among the weather forecasters for D-day—and lo and behold, they found one. No one but Irving Krick would claim such singular credit for meteorology's performance during World War II. Magazine writers and a biographer would uncritically fawn over the maestro meteorologist and embroider his tales of D-day forecasting exploits, and millions of readers would believe them. There were those who would not believe, of course, and would not forget. Stagg would not forget, nor Petterssen, nor Yates. Yates recalled in an official U.S. Air Force interview in 1980:

> Krick made life awfully difficult during the critical period. Krick and Petterssen, two prima donnas, got into an awful hassle. They completely disagreed. . . . I had to just ignore Krick's forecast the day before in order not to come out with a bad forecast. We had to throw it out. He said, "It's going to be good." He swears that he forecast the other way. . . .We [Yates and Stagg] could not understand . . . nor could any of the others on the phone understand the position . . . taken by Krick. Later, I figured out what he did. He got stuck with the system of analogs where he found a historical analogous situation that came out good. But boy, he had to be awful lucky and have a big mass of cold air come down from the top to clean it out, and there wasn't any mass of cold air building up up there. There was just nothing from any of the reports we had, and it didn't look the same. . . . But this was what he was basing his forecast on, and it was wrong. Although his historical analogue was a pretty good analogue right up to that time, . . . it blew, blew completely. The storms kept coming, and it didn't clear out. We got one good day, and that was enough to get us in there.

The Germans were surprised on June 6. The storms on June 4 and June 5 had grounded Luftwaffe reconnaissance flights and kept German naval patrol boats in harbor. Having been on alert throughout the fair weather month of May, the Nazi commanders looked on the stormy days of early June as a safe time for rest and relaxation. Their weathermen had very little data on conditions west of Britain, although within hours of Stagg's briefing of Eisenhower in the early hours of June 5, one German forecaster saw the same temporary break in conditions for June 6. By then, however, the key commanders, including Field Marshal Erwin Rommel, were away from their posts.

After the war, in 1948, Don Yates was in need of an eminent scientist. As commander of the U.S. Air Force Air Weather Service, Yates, now a brigadier general, created an important new Directorate of Scientific Services at AWS headquarters at Andrews Air Force Base in Maryland. He wanted AWS to be involved in pushing back the frontiers of weather science. He went looking for the best and most prestigious scientist he could find, and he got lucky. The first chief scientist for the Air Weather Service was the chief of the Norwegian forecasting service, Sverre Petterssen.

Meanwhile, a new administration at the California Institute of Technology took a new look at the cozy arrangements between its meteorology program and Irving Krick's private business interests. It didn't like what it saw. Caltech disbanded its meteorology program and severed all ties with its erstwhile chairman. Krick, meanwhile, found an increasingly profitable market for his brand of weather science salesmanship, which became more and more extravagant over the years. He could forecast the weather details of a single day a whole year in advance. He could make rain. When it came to weather, with a singular exception, there didn't seem to be anything that Irving Krick couldn't do. As it happened, what he could not do was remain a member of the American Meteorological Society. The organization's governing council informed Krick in 1958 that he had violated the AMS Code of Ethics and accepted his resignation. The president of the AMS at the time, although he took no part in those proceedings, was Sverre Petterssen.

PART V

Suddenly New Science

•

If ever there was a science remade, it was meteorology after World War II. The advent of the digital electronic computer put in the hands of both weather forecasters and atmospheric researchers a tool so powerful that the ways of their work changed suddenly and fundamentally. Born in the United States, Numerical Weather Prediction quickly spread around the world—and everywhere it went, it changed practically everything. It changed the way forecasts were produced and the way experiments were conducted. There is meteorology before Numerical Weather Prediction, and there is a different meteorology after Numerical Weather Prediction.

The key to it all was speed—not some special artificial intelligence brought to the science by the sophisticated new devices, but the brute speed of calculation. It was just as Vilhelm Bjerknes thought it would be, just as Lewis Fry Richardson dreamed: computers gave to meteorology the calculational capacity to become a thoroughly physical science. And so it did.

24

Jule Gregory Charney

Mastering the Math

•

"HARRY, IT'S SNOWING like hell in Washington," the meteorologist told the director of research for the U.S. Weather Bureau. The telephone call from Jule Charney came in the middle of the night. Charney was in his office in Princeton, New Jersey, wide awake, as he often was at that hour, and exuberant with his news about the weather in Washington, D.C., on November 6, 1952. At the other end of the line, Harry Wexler was at home in Washington, in bed, formerly sound asleep. This urgent weather report came not in November 1952, but a year later, in 1953, although more remarkable at the time was the fact that it came at all. The storm of November 6, 1952, which dumped six inches of snow on the nation's capital, had come as a surprise to weather forecasters in 1952. But now a powerful new tool was about to change meteorology. Charney and his elite team of researchers at the Institute for Advanced Study had programmed the same data used by the mistaken forecasters into their unique mathematical model running on a newfangled electronic computer. Charney looked at the results of the test forecast: the storm that had fooled the forecasters had not fooled the computer. He just couldn't wait until morning to break the news to Wexler. It was snowing like hell in Washington, and history was being made.

Through the first half of the twentieth century, the best description of what meteorology really needed still was Lewis Fry Richardson's fantastical vision of a "weather forecasting factory," a great, cavernous hall occupied by tens of thousands of individual calculators, each responsible for a certain parcel of atmosphere. Forecasting practices had improved since the early part of the century, of course, but not enough. While important new conceptions of the structure and origin of storms had migrated through the science, these advances did not translate very directly into more accurate weather forecasts. Important new details had been added, telling airline pilots what visibility and upper-air conditions to expect, and precipitation forecasts had become more reliable. Still, a moment would come when forecasters would have to fall back on a certain amount of guesswork—however well-educated, however scientific—what theorist Carl-Gustaf Rossby called "the horrible subjectivity." Vilhelm Bjerknes's dream in 1904 of rendering meteorology an exact science, of calculating changes of the weather according to the laws of physics, seemed as distant as ever. The physical laws had been known for years, but the old problems remained: the behavior of the atmosphere was just too complicated, and the calculational effort was just too big. What meteorology needed, of course, was a fantastically powerful new way of handling its voluminous data and the incredible mathematical detail of its turbulent fluid dynamics. Until the end of World War II, however, no such thing was in sight.

The turning point came with the advent of the electronic digital computer and mathematician John von Neumann's decision in 1946 to try to demonstrate the social and scientific potential of such a device by putting it to work on the weather. Like the thermonuclear explosions that were his first objects of study, the atmosphere appeared to von Neumann to present the kind of especially difficult problems in nonlinear fluid turbulence he thought the computer would be uniquely able to solve. His interest in the subject was cultivated by conversations with Vladimir Zworykin, an instrument engineer at Radio Corporation of America; with Rossby at the University of Chicago; and Francis Reichelderfer at the Weather Bureau. In fact, it was not so much weather as warfare that always would attract von Neumann's visionary genius. To his rabidly hawkish frame of mind, if weather could be accurately simulated, perhaps it could be controlled, and if controlled, perhaps used as a weapon. However ulterior his motives, his plans for the new digital electronic computer under development at the Institute for Advanced Study in Princeton, New Jersey, would forever change the science of meteorology.

The importance of the Meteorology Project was not obvious in 1946, and several years would elapse before the method that became known as Numerical Weather Prediction would prove itself. For years, practicing weather forecasters harbored fears that they would be replaced by machines. Well into the 1950s, more fundamental scientific doubts would linger. "It is bad technique to apply the sort of scientific method which belongs to the precise, smooth operations of astronomy or ballistics to a science in which the statistics of errors is wide and the precision of observations is small," declared mathematician Norbert Weiner at the Massachusetts Institute of Technology. "In the semi-exact sciences, in which observations have this character, the technique must be more explicitly statistical and less dynamic than in astronomy."

Describing his plans in an August 1946 meeting with leading meteorologists, von Neumann was not surprised by the halfhearted reaction from the group. He and Rossby already had concluded that meteorological theory was not ready for digital computing. "Indeed, the possibilities that are opened up by these devices are so radically new and unexpected, that the theory is entirely unprepared for them," von Neumann had written before the meeting. "There was no practical motivation in the past to work out those parts of meteorological theory on a mathematical and analytical level, which in order to become really effective, would require calculational methods that are 1,000 to 100,000 times faster than what seemed possible at the time!" A conference of meteorologists in December concluded that meteorology was not yet ripe for this change.

Rewriting meteorological theory for the sake of experimental digital computation was not an attractive proposition to most meteorologists. What assurances were there that Numerical Weather Prediction by such a untested device would accurately forecast weather? The only certainty was that the effort would be incredibly difficult mathematically and theoretically, demanding a level of scientific talent that few meteorologists possessed. Besides the visionary Rossby, who would be willing to risk a reputation and invest a career in such a venture? The only attempt ever made, the arduous hand calculations made by Richardson in 1920 during World War I, had ended in internationally famous failure.

Among leading meteorologists, expectations were low, not only for Numerical Weather Prediction but for improving forecasting accuracy by any other means. Rossby's successor as president of the American Meteorological Society, Professor Henry G. Houghton of MIT, told the organization in December 1946, "There appears to be no immediate prospect of an objective method of forecasting based entirely on sound physical principles." While current subjective methods might be improved somewhat, he said, "it must always be remembered that such efforts are in the nature of palliatives rather than cures." As late as 1952, two years after the first successful test of forecasting by computer, C. K. M. Douglas of the British Meteorological Office proved himself better at forecasting English weather than the future of his science. In the *Quarterly Journal of the Royal Meteorological Society,* he wrote, "On the whole the prospects of computing the future weather, with or without electronic machines, look remote at present." Douglas was a veteran synoptic meteorologist, an intuitive forecaster in a science that was just then being reborn. Certainly, he was not very closely following events taking place around Jule Gregory Charney in those days or taking account of the brilliant intensity of the young American.

Jule Charney was a Californian, born in San Francisco on New Year's Day 1917 and raised since the age of five in Los Angeles. He was the son of Stella and Ely Charney, Yiddish-speaking Russian immigrant Jews. They worked in the garment industry; his mother was a seamstress for movie studios for a time. It was an intellectually stimulating and somewhat volatile home life, enlivened with strongly held left-wing political views, and enriched with Stella's musical talent at the piano and with strong family links to writers, scholars, and artists. A medical diagnosis of heart trouble, which eventually proved mistaken, left Jule with the feeling that he was physically inferior to other children. Expectations for Jule were great, and by the middle years at Hollywood High School he discovered his special gift. His recollections came during an extended interview with George W. Platzman in 1980 that was published as part of a memorial volume, *The Atmosphere—A Challenge*. "I was a very shy, retiring kid [who] didn't have many friends," he recalled, and that was a troubled time. Amid the normal emotional turbulence of adolescence, his parents separated, and his home life fell apart. At an uncle's house during a miserable temporary stay in New York, he happened upon a classic calculus text. "And just to wile away the time, I remember reading the introductory part, and the first part of it, and finding that it was very comprehensible, and that I could do the problems." Mathematics became a consuming passion, "because I could do it! I suddenly discovered that I could do it." At a time when no one in high school and few in college were being taught calculus and differential equations, Jule at the age of 15 was teaching himself. Jule remembered focusing on a particular mathematical text from the public library. "I went through the whole damn thing, differential and integral calculus, doing all the examples, religiously," he recalled. "And then I got a book on differential equations . . . and I found that it was duck soup."

The University of California's Los Angeles campus was a young school that was notably weak in mathematics and physics and might have been an unfortunate choice for a young man of Jule Charney's prodigious analytical talents. But it was close to home, the country was in the throes of the depression, and no one ever advised the young man that he should consider a more established university. "I was really interested in theoretical physics, and there were no theoreticians," Charney would recall. Unchallenged, uninterested, he fell into bad habits. "I didn't have to study at all," he recalled. "Nobody encouraged me." As it happened, those unsatisfactory circumstances at UCLA conspired to attract Charney to meteorology.

Jule seemed to be on his way to a career in mathematics or physics, obtaining a bachelor's degree in 1938 and a master's in 1940, before he first encountered the science of weather. He attended a seminar lecture by Jörgen Holmboe, a young Norwegian professor from the Geophysics Institute in Bergen, who first exposed Charney to the idea that motions of the atmosphere are subject to the laws of physics and can be described by partial differential equations. It was not love at first sight. "I wasn't terribly impressed by it, but maybe impressed enough to say, 'Well, here meteorology is a fairly serious science.' Up to that point, I knew zero about meteorology."

Young Holmboe, who was part of the new UCLA meteorology program established by Jacob Bjerknes, invited Charney to become his teaching assistant. The war was enlarging in Europe, and UCLA's program was part of a national campaign sponsored by the army and the navy to provide weather forecasting training to thousands of servicemen who would be needed in the increasingly likely event that the United States entered the war. Young men in Charney's position faced serious choices. Some were joining the armed services; others were accepting defense-related positions, which, like Holmboe's teaching assistantship, offered deferment from the military draft. After considering aeronautics, Charney chose meteorology and accepted Holmboe's offer. During the war, Charney would help train scores of wartime meteorologists, although he would later joke about how little he knew of the subject when he enrolled on July 1, 1941, as a student in class number two. "I came to meteorology from mathematics and physics with the suspicion but not the certainty that wind was moving air," he wrote. "I had never seen a weather map. Learning meteorology was for me a slow and painful process. I didn't like it. The atmosphere did not fit into the neat and simple categories I had come to expect from mathematics and theoretical physics."

However slow and painful, meteorology at UCLA in those years posed a level of intellectual challenge to Charney that mathematics and physics did not. An outstanding meteorology program had sprung into existence with the coming of Bjerknes and Holmboe from Norway and the German-born Morris Neiberger from MIT. Charney became an accomplished instructor in map analysis and forecasting, although always his natural interest was theory. And his friend Morton G. Wurtele would remember those years as a time of great social development for Jule. "He was fun-loving, always ready for a joke, a laugh, a prank, verging on the lecherous, arguing politics at the drop of a slogan, eager to participate in, but without any special talent for, whatever athletic activity anyone proposed," Wurtele wrote. His students stuffed the ballot boxes to ensure his election as "Professional King" of the Mardi Gras. "Thank you for the honor of this role," remarked Charney, "but where is my queen?" In 1944, friends introduced Charney to Elinor Kesting Frye, a graduate student in philosophy. They were married in 1946 and had two children in addition to a young son of Elinor's by a previous marriage.

As the fortunes of war turned in favor of the Allies and Charney's teaching chores eased, he returned to his studies as a graduate student. It was time to get serious about his own potential as a scientist. Was he capable of serious research in meteorology? In 1946, it was time to find out. Already he knew his subject. He was going to sink his teeth into an important theoretical question that meteorologists had been struggling with for years. How do the big storms of the midlatitudes originate? The Bergen school's polar front model illustrated the development and progress of these cyclones, but the mechanism of their origin remained a mystery. A 1944 paper by Bjerknes and Holmboe pointed to upper-atmosphere waves and divergent airflows, but it all began at the surface of the front. To Rossby in Chicago, the center of action was in the middle atmosphere at the level of the planetary waves he had defined in his famous 1939 paper. The three Scandinavians reached different conclusions using different models and different analytical methods. Adopting Bjerknes's three-dimensional depiction and a level of analytical treatment more typical of Rossby, Charney stepped right into the middle of this controversy and settled it.

Charney's doctoral thesis, "Dynamics of Long Waves in a Baroclinic Westerly Current," was a scientific sensation. In a rigorously mathematical treatment of the subject, he formulated an important concept of modern meteorology. Storms of the middle latitudes form in response not to instabilities along the front, but to perturbations caused by a larger atmospheric circumstance. Cells of circulation break down into big eddies when the atmosphere is characterized by sharp horizontal temperature differences and vertical, westerly wind that gains speed with height. This condition, known as *baroclinic instability,* is most commonly evident during the sharp temperature contrasts of midlatitude winters. Its consequence is the large-scale pattern of rising warm air in low-pressure cyclones and falling cold air of high-pressure anticyclones.

Arriving at this conclusion, Charney devised certain critical mathematical "approximations" that exhibited what one leading meteorologist has called "physical insight amounting to genius." These shortcuts were just the kind of approach that would be needed for converting a mathematical model of the atmosphere for use by an electronic digital computer.

Charney's thesis was published in the October 1947 issue of the American Meteorological Society's *Journal of Meteorology,* which was founded by Rossby; it took up the entire issue. "It was a totally original work," recalled Morton Wurtele, who went on to become chairman of meteorology at UCLA. "Even though the basic concept is easy to understand, it was very mathematical, and neither Bjerknes or Holmboe were trained in mathematics the way Charney was, so you might say Charney did his thesis without important guidance from anybody." Meteorologist Norman A. Phillips, himself a pioneering member of the Princeton team, wrote of Charney's thesis: "For the first time, this paper established a believable mechanism for the development and motion of the large-scale disturbances in the atmosphere."

In 1946, Charney accepted a National Research Council fellowship to study a year in Norway. On the way, while Elinor visited family in the Midwest, Jule stopped at the University of Chicago to meet Rossby. Before he finally pulled himself away from Chicago, nine happy months had breezed by. Charney later called those months "the main formative experience of my whole professional life." In Rossby he found a kindred spirit, another brilliant intuitive theoretician of the same analytical bent. "We thought very much the same way," he recalled, "and we had *endless* conversations. . . . There was no subject that we couldn't discuss and it was always intensely interesting to discuss things with him." The conversations continued through the cocktail lounges and restaurants of Chicago. "It was just enormous fun . . . one of the most delightful, the most exciting periods of my whole life," Charney recalled. "I discovered what it meant to have intellectual rapport with a man . . . with another person." In August, with Rossby, he was among 18 meteorologists who attended von Neumann's Meteorology Project meeting at the Institute for Advanced Study in Princeton. There were those who attended who came to think of this as the most important outcome of the meeting—getting Charney and von Neumann together.

Charney went off to Norway in the spring of 1947, focusing his thinking on a way to further simplify or filter out the irrelevant "noise" in the difficult mathematical models of the atmosphere. Rossby, meanwhile, after more than 20 years in the United States, was making plans to return to Sweden, where he would organize an international meteorological institute. At the Institute for Advanced Study, von Neumann struggled to recruit a qualified staff for the Meteorology Project. Air Force Lieutenant Philip Thompson, a good friend of Charney's from UCLA, had been hired by von Neumann, but the work in Princeton was getting off to a slow start. In fact, von Neumann considered dropping the whole idea. Practically alone in Princeton, Thompson wrote a long, frustrated letter to his friend in Oslo, posing a number of questions about various waves in the atmosphere. In February 1947, Charney wrote a long, commiserating reply, offering serious technical advice as well as creatively playful metaphor. He wrote:

> We might say that the atmosphere is a musical instrument on which one can play many tunes. High notes are sound waves, low notes are long inertial waves, and nature is a musician more of the Beethoven than of the Chopin type. He much prefers the low notes and only occasionally plays arpeggios in the treble and then only with a light hand. The oceans and the continents are the elephants in Saint-Saens' animal suite, marching in a slow cumbrous rhythm, one step every day or so. Of course, there are overtones: sound waves, billow clouds (gravity waves), inertial oscillations, etc., but these are unimportant and are heard only at N.Y.U. and M.I.T.

Soon after came word from Oslo of another theoretical advance by Charney. Building on his thesis, he had more completely formulated a mathematical shortcut called the *quasi-geostrophic* approximation. This formula substituted for the actual winds in mathematical models the standard values of the systematic westerlies that are formed by a balancing of pressure differences against the accelerations caused by Earth's rotation. Using this approximation filters out lengths of atmospheric waves that are not important to meteorology, Charney argued. The quasi-geostrophic model reduced six equations of atmospheric motions to a single equation that described those motions of the atmosphere that are most important to weather. The concept, important for theory, was crucial to the success of Numerical Weather Prediction.

In a letter to Thompson written from Oslo on November 4, 1947, Charney said he had been "brooding about" the electronic computer project since arriving in Norway. "The more I see of the feeble efforts of meteorologists here and elsewhere, including myself, the more I am convinced that weather forecasting is a computing problem, and that its solution requires one highly intelligent machine and a few mathematico-meteorological oilers," Charney wrote. "Well at present it is my impression that you have, or are about to get the machine, but that sufficient oilers are lacking."

In the summer of 1948, having returned from Oslo, Charney was hired by von Neumann to work on the Meteorology Project and immediately took over as mathematico-meteorological oiler in chief. The Meteorology Project—near extinction in early 1948—became under Charney's leadership highly organized and brilliantly staffed. From Stockholm, Rossby followed Charney's progress with intense interest, urging "a systematic test and extension of Charney's method." In a May 1949 letter to meteorologist George W. Platzman, a Princeton team member, Rossby wrote, "I must confess that I have an extremely strong feeling that we are standing at the threshold of a new era in applied meteorology and that we must push this line to the point where it can be put in general operation." In the spring of 1950, because the institute's computer was not yet complete, Charney's team first successfully demonstrated Numerical Weather Prediction on the ENIAC computer at the Army's Ballistic Research Laboratory in Aberdeen, Maryland. "All in all," Charney wrote at the time, "I think we have enough evidence now to bear out most of Rossby's prophecy. Of course we shall want more." The Princeton team estimated that it had taken 24 hours to compute a 24-hour forecast, but jokes about that improbable pace wouldn't last long. By the spring of 1952, the new and faster computer at the Institute for Advanced Study finally was up and running. Progress was swift and dramatic. Before long, it was snowing like hell in Washington. On May 6, 1955, a new era in weather prediction began with real-time daily forecasting by a unit composed of U.S. Weather Bureau, Army, and Navy personnel on an IBM 701 computer.

Before long, the output of computer models would be the technical starting point for virtually all public and private weather forecasting around the world. The formulation of improved and more sophisticated mathematical models on ever-faster computers would follow from a steady stream of fruitful research. But the progress would become increasingly incremental, and the process increasingly collective. Nothing again would compare to the defining moments of 1946 when John von Neumann decided to use weather as the first example of what the electronic digital computer could do for society, and when Jule Gregory Charney, more than anyone, made it happen.

"This was a unique opportunity for meteorology, the equivalent of which may never occur again," Norman Phillips would recall. "I am convinced that without Jule's willingness and ability to study the hydrodynamical aspects, this first attempt would have failed. The ingenious quasi-geostrophic system might have become a theoretical curiosity for a long time, with little hope for thorough experimental verification. I believe this aspect of Jule—the importance he attached to comparing theory with reality, and his awareness of the care that this often requires—is something we should be even more grateful for then we are for his well-known theoretical ability."

Building on Charney's research, Phillips in 1955 devised a successful mathematical model that allowed the Institute for Advanced Study's computer to simulate accurately the general circulation of the atmosphere. It was the first successful climate model, describing statistically consistent monthly and seasonal patterns of the atmosphere. The advance by Phillips illustrated that the computer would be as powerful a tool for meteorological research as it would for weather forecasting. Charney hailed it as a breakthrough, calling the model the beginning of the science of "dynamic climatology." Phillips's was the first of a long line of general circulation models that would be the basis for studies into the potential climatic effects of increasing concentrations of industrial carbon dioxide and other so-called greenhouse gases.

Their mission accomplished in Princeton, the scientists of the Meteorology Project left in the mid-1950s and headed off to university and other positions. Charney went to MIT in the summer of 1956. There, in the tradition of Rossby, who suddenly died of a heart attack a year later in Stockholm, Charney for 25 years inspired and guided the development of a generation of leading atmospheric scientists. He was 64 years old when he died of cancer in Boston on June 16, 1981.

25

Jerome Namias

The Long Ranger

•

FOR A FELLOW WHO considered himself lucky, the man who pioneered the modern science of long-range weather prediction had more going against him than anyone since the nineteenth-century theorist William Ferrel, who had to carve the geometries of the atmosphere into the barn door with a pitchfork. Jerome Namias was a special case: a leading scientist in an increasingly rigorous discipline who never earned a doctorate or even a bachelor's degree. Recognition would come with numerous scientific awards and even, to his amazement, his election to the National Academy of Sciences. He was lucky to have been in the right place and the right time, he would explain, but others would see him differently. Jerome Namias's idea of good luck was overcoming the kind of bad luck that could have been devastating to a less ardently determined and brilliant individual.

Certainly he was lucky in the late 1920s to have had an excellent high school physics teacher in Fall River, Massachusetts, who inspired his ambition to become a meteorologist. In an autobiographical sketch published by Scripps Institution of Oceanography in 1986, Namias recalled the misfortunes of his youth without a trace of self-pity or regret. Just as he was offered a four-year scholarship to Wesleyan University, his father, an optometrist for the textile workers of Fall River, became seriously ill. Because his older brother, Foster, had already moved away to New York, he would recall, "I felt my place was at home to help out our mother who had always done so much for the family." In what would have been his freshman year at the Connecticut university, he was selling sundry items door to door in Fall River and earning a few dollars as a drummer in a jazz band. Near the end of the year, as his father regained his health, Jerome Namias came down with tuberculosis. While he had to "shelve my idea of going to college," he recalled, the long convalescence and confinement gave him a lot of time to study at home. Physics and mathematics and meteorology came through correspondence courses, along with German and English composition and chemistry. Finally back on his feet, he found himself looking for a job in weather science, as luck would have it, during the worst days of the Great Depression. He wrote hopeful, polite letters to meteorologists who had written papers he had read, but he had to admit that prospects of employment were not very bright: "Who wanted a student meteorologist, especially one with no formal qualifications?"

The first break came in 1930 by way of a letter from Henry Helm Clayton, a research meteorologist in Boston, who invited Namias to his home near the Blue Hill Observatory. Clayton explained that he was doing research with Charles G. Abbot, secretary of the Smithsonian Institution, on the influence of solar radiation on weather. Their data, which would be published by the Smithsonian as part of a climate data series called the World Weather Records, was buried in the archives of the Weather Bureau's central office in Washington, and Clayton hired young Namias to go dig it out. Namias became an employee of the Smithsonian under Abbot. Most important, he remembered, the job gave him access to the Weather Bureau's library. There he encountered the recent breakthrough papers of the Bergen school meteorologists in Norway, which Weather Bureau forecasters were so assiduously avoiding at the time. And he found and read with great interest, and some difficulty, the first publications from Carl-Gustaf Rossby's new Meteorology Department at the Massachusetts Institute of Technology.

The second and biggest break, what Namias would always regard as the defining event in his life, came in 1931 from Rossby. Young Namias had read a paper by the Swedish theorist and was uncertain about two points raised by the scientist. In a politely worded letter, he questioned the accuracy of these two assertions. Soon came a reply from Rossby saying that Namias was correct on one point and explaining at some length why he was wrong on the other. By way of a postscript, Rossby wanted to know more about his correspondent and invited him to visit the next time he was in Cambridge. "I needed no urging," Namias recalled, "and in a month or so, had the pleasure and opportunity of meeting the person who was to influence my entire life thereafter." Always scouting for talent, Rossby immediately recognized in Namias an uncommon intelligence and a practical gift for meteorology. Rossby invited Namias to study at MIT. "With my unorthodox background, Rossby had to make a 'special case' of me," Namias said. Under Rossby's big wing, Namias found both confirmation of his future as a meteorologist and the financial security, such as it was, of various part-time jobs. More than that, like many weather scientists of his generation, he found a leader in Rossby. "He set me on a path of excitement and gave me a philosophy that led to achievement," Namias would recall, "and equally important, set my standards of behavior and ethics."

Namias quickly was recognized as an outstanding synoptic forecaster, a master of the map, and a gifted and eager student of the Bergen school methods of air-mass analysis. Working side by side with the best research meteorologists in the world, Namias always would be a special case, employing different methods and moving in different directions. He was a researcher with little formal education in a time and a place where meteorology was becoming increasingly specialized and technical. He was an empiricist in an age of theory, devoted to observational meteorology in an era when many leading weather scientists were focusing on more formal dynamical studies and the mathematical rigors of Numerical Weather Prediction. But his talents were well known and well regarded. After World War II, when Jule Charney and his team were testing their numerical models on the first computer, it was Jerome Namias who would be called in to apply the ultimate test: "to see if the computer-generated forecasts resembled the real atmosphere."

Namias's renown as a scientist came with his research into ways of extending predictions beyond the conventional range of what at the time was one or two days. This professional path was set in the 1930s by the climate calamity known as the Dust Bowl. With the desperate flight of tens of thousands of impoverished farmers from the Great Plains came the first federal funds for research into long-range weather prediction. It was this funding that allowed Rossby in 1936 to offer Namias a full-time position as a research assistant at MIT. "Nothing could have made me happier," he recalled, "for I was to embark on my truly professional career."

In the late 1930s, Rossby was developing his theory of planetary waves, the enormous oscillations in the upper atmosphere that came to be called *Rossby waves.* These big patterns in the westerly winds had been emerging from an increasingly widespread network of direct upper-air observations from pilot balloons and aircraft. At East Boston Airport, Namias had been given the job of collecting data from the daily flights of a weather observation plane dedicated to MIT's program. Rossby was building on the findings of the Norwegians, whose dense network of surface observations had led to the description of the structure of individual storms. Taking advantage of the continental expanse of North America, the MIT team was using their new aerial data to grasp the very large scale of weather and to better describe its three-dimensional structure. Still, the upper waves in the westerlies were so global in scope that only partial waveforms were revealed by the network of observations existing in 1940. Frustrated, Rossby asked Namias to see if he could extend the map of upper-air data beyond the continental limits of the United States. Namias set to work on surface data from ships in the Pacific and the Atlantic, employing the kind of differential analysis first explored by LeRoy Meisinger in the 1920s to extrapolate conditions in the air aloft. The large-scale pattern emerged on the map he drew.

"When Rossby saw it, he got very excited," Namias recalled in an interview with science writer Janet Else Basu in 1984. "The long waves were vivid—instead of having just one of these troughs, there was one over the Atlantic and one over the Pacific. Not even bothering to find a piece of paper, he set to work scribbling right on the map to compute the motion of these waves. . . . Those first computations on the side of the map triggered a whole new chain of thinking in the 1940s. They formed the basis for the advance in extended forecasting." Rossby's formulation for the slow movement of the large waves opened the way for experimental five-day forecasting by Namias and others in the MIT program.

Just as the looming entry of the United States into World War II gave urgency to the training of military meteorologists at key universities across the country, so did it hasten the extended-range forecast experiment at MIT. In 1941, at the urging of military leaders, MIT's program was transferred to the Weather Bureau in Washington. Namias was granted a one-year leave of absence to become chief of the agency's new Extended Forecast Division.

Namias's Weather Bureau team provided five-day forecasts for the army and the navy throughout the war. He lectured to training courses for military meteorologists and conducted several studies in connection with military operations. "Our big success story was the forecast for the Allied landing in North Africa on 8 November 1942. It was very important that we got the state of sea right, and for this we used techniques developed by [Harald] Sverdrup and [Walter] Munk that depended on estimates of the wind fields over the North Atlantic for several days previously." For this work he earned a citation from Secretary of the Navy Frank Knox for making "an important contribution to operations which resulted in the successful completion of landing operations by the largest landing force in history."

For 30 years, until 1971, Namias would remain at the Weather Bureau as chief of the Extended Forecast Section. In 1950, the Weather

Bureau began issuing official 30-day weather outlooks. In 1958, 90-day forecasts began. Namias developed a variety of techniques that would help him try to forecast general weather conditions beyond the range of conventional methods, and he conducted several important diagnostic studies of weather events. "Like so much of his research, these empirical studies, and particularly the conceptual framework in which he interpreted the results, were often decades ahead of their time," Eugene M. Rasmusson, a noted research meteorologist at the University of Maryland, wrote in a tribute in 1998. "After attracting initial interest, they often faded into the background, only to be fully appreciated a generation later as additional empirical, modeling, and theoretical work revealed their fundamental significance." One landmark paper described variations in the jet stream that affect vertical motions in the atmosphere. Another identified an "index cycle" of seasonal changes in the westerlies which, under another name, the Arctic Oscillation, is the subject of cutting-edge climate research at the turn of the new century.

Publicly at least, Namias stayed clear of the controversies continually surrounding Irving P. Krick and his extravagant claims of long-range forecasting accuracy. More than once, however, he would make clear his dislike of the analog method so ardently promoted by Krick. "I have seen false hopes engendered and large amounts of money spent on the analog method, but with very little to show for it," he told an interviewer. "To be frank, the philosophy of analogs runs counter to mine because it short-circuits understanding." Often he warned potential users of long-range forecasts not to be "too gullible" in the face of extravagant claims. "Since meteorologists do not have a proven approach," he told one group of energy suppliers, "the door is open for charlatans who will forecast what is going to happen many years ahead."

Long-range weather prediction would remain the subject of skepticism and controversy among meteorologists throughout the second half of the twentieth century. In contrast to the rigorously mathematical- and physics-based work under way at the Numerical Weather Prediction projects, long-range forecasting was empirical and often subjective. For several decades, the field drew less support and scientific interest than the revolutionary development of computer model forecasting. In the hands of a lesser scientist, such an experimental scientific research program in such a bureaucratic environment might have collapsed altogether. Namias's classical scientific approach, the quality of his numerous case studies and other scientific papers, and his stellar pedigree as a Rossby protégé combined to earn the field a high level of respectability.

Beginning with his studies of the index cycle, Namias in the 1950s focused on what he called "long-period evolutions" of atmospheric systems. For answers to unusual seasonal conditions, he looked to the slow migration of pressure regimes and the movement of the Rossby waves in the upper atmosphere. These large atmospheric shifts helped explain periods of summertime drought experienced in the West and unusual winter weather patterns.

His research took a major turn in 1958 after a conference in La Jolla, California, held by the Scripps Institution of Oceanography, that explored the unusual Pacific Ocean conditions of 1957–1958. This was a period of intense global research known as the International Geophysical Year, which coincidentally turned out to be a time of unusually warm surface temperatures across the tropical Pacific. In a series of papers in the 1960s, Namias was first to shift the focus of long-range research from the atmosphere to the interplay between the atmosphere and the surfaces it encounters across Earth. This idea had been suggested years earlier by Scandinavian scientists, but only in the 1960s, because of the global reach of the multinational IGY effort, did researchers have the observations that could test these effects. Namias called attention to the effects on the atmosphere of ice and snowfields, the soil with its different moisture characteristics, and, most important, the surface temperatures of the ocean. "The well-mixed layer of the ocean retains its thermal signature down to a few hundred meters," he noted. "Since water has a high heat capacity, the first three meters of the ocean contain as much heat as the entire atmosphere above." At UCLA during this time, Jacob Bjerknes was using the same data from the International Geophysical Year to link the unusual ocean conditions to Sir Gilbert Walker's turn-of-the-century Southern Oscillation statistics, and so to define the tropical climate anomaly known as El Niño.

By 1971, Namias was ready for a change. After 30 years, he had grown weary of budgetary and political battles in the Washington bureaucracy, and his case of Potomac fever finally had run its course. Rossby had died of a heart attack in 1957. His best friend and brother-in-law, Harry Wexler, director of research for the Weather Bureau, had died of heart trouble in 1962. Namias himself had suffered a heart attack in 1963, and in 1964 he was seriously injured in an automobile accident south of Boston while returning from the Woods Hole Oceanographic Institution. His longtime research collaborator, Phil Clapp, was retiring from the newly renamed National Weather Service, so in 1971 Namias and Clapp decided to collaborate one more time—on a retirement party.

For Jerome Namias, retirement meant another invigorating career as a research scientist. He accepted an offer to begin the first program of climate research studies at Scripps in La Jolla. With new enthusiasm, he plunged into further research on the interplay between the upper levels of the ocean and the lower levels of the atmosphere.

In 1974, as millions of American consumers were feeling the impact of an oil embargo by Arab nations, the Carter administration was contemplating the politically onerous expedient of issuing gasoline rationing cards to Americans. High government officials called on Namias's expertise as a long-range forecaster. His prediction that the winter would be warmer than normal encouraged the government to forestall the implementation of fuel rationing.

During his years at Scripps, Namias published some 70 papers, including some work on El Niño. Namias was first to identify the cold, opposite extreme of tropical Pacific sea-surface temperatures, the condition now called La Niña. Still, he remained generally unconvinced of the growing recognition among his colleagues that conditions in the Tropics often lead to changes in midlatitude weather and climate variations. This resistance on his part seems all the more curious for his close association and personal friendship during these years with the leader of this school of thought, the great Jacob Bjerknes. Maybe this was a trick the old dog just couldn't quite master. For 50 years, he had been looking at Weather Bureau maps of the Northern Hemisphere centered on the pole, focused on the migration of Rossby waves among other features, and data from below 20 degrees north was generally nonexistent during this epoch. Namias was not alone among Northern Hemisphere specialists of this time who were inclined to think that conditions in the middle latitudes more likely forced variations in the Tropics, rather than the other way around.

Nobody pressed him too hard on this point at this stage of his long and fruitful career. He was revered and beloved. Rasmusson recalled the occasion of Namias's 75th birthday, which was marked by a special symposium in his honor, and the words Alan Hecht used to pay tribute to his contributions. "If you could look up 'Namias' in the dictionary you would find the following definition: 'a man who gives good reasons for any long range forecast, and even better reasons for why it fails.' This is the character of a man who is an infinite source of good ideas, and who thinks fast on his feet." In the late 1970s and early 1980s, Namias was a research meteorologist without peer, the dean of climate research, one of the last in a line of real giants in the field of meteorology. At the age of 79, Jerome Namias suffered a stroke, and never fully recovered. He died on February 10, 1997.

26

Edward N. Lorenz

Calculating Chaos

•

THE POSTWAR REVOLUTION in meteorology that was ignited by digital electronic computers gave rise to a long dawn of dreamy ambition among weather forecasters and researchers. In the 1950s and 1960s, the study of weather was finally coming into its own as both a science and a service. At last, scientists had the technical wherewithal to apply the objective tools of the laws of physics to unraveling the old mysteries of the atmosphere. Those ingenious Numerical Weather Prediction computer programs were instruments of incredible power and scope, and the quickening pace of technological advance seemed to promise limitless potential. The whole turbulent, unwieldy problem had been compressed into something like an electronic bottle and at long last brought into the laboratory, where hypotheses could be tested and where increasingly sophisticated and accurate model atmospheres could bring every last detail into finer and finer focus. Not only had usable weather forecasts been achieved, but a workable model climate had been created.

It was a boom time. Returning World War II veterans had peopled a population explosion in the science at universities, in the U.S. Weather Bureau, and in private industry. Governments were supplying increasing funds for its growth, fueling major research programs. The combatant nations found themselves with large air forces requiring elaborate meteorological services, and growing civilian aviation fleets were making increasingly challenging demands on weathermen. Technology was defining both the problem and the solution. Advances in computing power, rapidly enlarging networks of observations, and the successful launch of TIROS I, the world's first weather satellite, all inspired confidence that meteorologists could accomplish their new mission. Researchers and forecasters who long had wrestled with the mysteries of weather seemed free at last from the hoary veil of frustration and disappointment.

From long labor was coming the fruits of an old promise. In 1814, the great French mathematician Pierre-Simon Laplace had famously described a vision of a clockwork world that physical scientists had long taken as an article of faith. The present state of the universe, wrote Laplace, should be seen "as the effect of its prior state and as the cause of the one that will follow." He envisioned an "intelligence which at a given instant knew all of the forces by which nature is animated and the respective situation of the things of which nature is composed . . . nothing for it would be uncertain, and the future, like the past, would be present to its eyes." Laplace had inspired Vilhelm Bjerknes's call a century later for a rigorously scientific meteorology; and as recently as 1959, the Swede Tor Bergeron described the weather forecast as "the most important and promising but still unsolved Laplacian problem on our planet."

Free and flush, impressed with Numerical Weather Prediction and infected with the optimistic spirit of the era, weather scientists in the 1960s greeted the new day with soaring expectations. Some allowed themselves to dream of a time when forecasting would finally attain the precisional heights of astronomy, its older and more venerated sister science. As in astronomy, the laws of Isaac Newton had finally taken their rightful place in meteorology, and now what remained for weather science was to achieve the same degree of exactitude. Like astronomers calculating the distant return of a comet, meteorologists could see themselves wielding powers of prediction farther and farther into the future.

More than that, the day was not far off when weather not only would be forecast to perfection, but its rains and winds would be tamed. Like the bounties of land and sea, the unruly products of the sky would be intelligently redesigned to the purposes of enlightened humanity. The dangers of fogs and hails would be relics of the past. Man-made showers would prevent droughts. Man-made droughts would prevent floods. Hurricanes would be suppressed or steered harmlessly out to sea. Barring enlightenment, John von Neumann's ambition was gaining new currency: weather control would become a weapon of war.

Even as weather scientists were chasing these goals, however, another line of research was beginning to cast a very different light on the character of the atmosphere and the problem of accurate forecasting. While most meteorologists concerned with weather prediction were trying to extend the range or improve the accuracy of forecasts, an uncommonly gifted researcher at the Massachusetts Institute of Technology was going his own way. Ed Lorenz was using his new Royal McBee computer to operate a stripped-down numerical forecasting model that posed this different, more fundamental question about the atmosphere: Just how predictable is it? In 1961 came an important answer that was a showstopping disappointment to the most optimistic weather forecasters.

In the hands of a less mathematically inclined scientist, the fundamental limits to the detailed predictability of the atmosphere might have gone undefined for years. In the event, it required a kind of intuitive leap that only a born mathematician was likely to make. "As a small child, I enjoyed playing with numbers," Ed Lorenz would recall of his days growing up in suburban West Hartford, Connecticut, as the son of a mechanical engineer in a home where the game of chess was popular recreation. He went to Dartmouth College, where he majored in mathematics, and then in 1938 he enrolled at Harvard University as a graduate student in math. Like Jule Charney, Lorenz was bred into a meteorologist in the 1940s by the exigencies of the impending war. He came down Massachusetts Avenue from Harvard to MIT to learn how to be a weatherman in one of the nine-month crash courses that trained thousands of military forecasters as the country entered World War II. He served in the Army Air Corps, forecasting conditions for bomber pilots out of weather centrals in Saipan and Okinawa. At the end of the war, after pondering his choices, he decided he could accomplish more for meteorology than mathematics and accepted a research position at MIT. He would study dynamics and teach statistical forecasting for a time, occupying a special disciplinary realm between research and application—not predicting weather but, more fundamentally, investigating its predictability.

Some statistical forecasters were arguing at the time that certain of their techniques could achieve the same accuracy as Numerical Weather Prediction. In the winter of 1961, Lorenz was putting that claim to the test, running a stripped-down model atmosphere on his Royal McBee LGP-30 computer. If patterns of periodic behavior emerged, he reasoned, the statisticians might be right. Certain patterns repeated themselves, just like weather, although they were not regularly periodic. Along the way, however, he encountered a circumstance that looked at first like an accident or a computer glitch.

He had interrupted the running of the program and wanted to restart it at a midway point in its operation. To do this, he retyped into the computer the last values of the simulated weather variables it had printed out. And then he went down the hall to get a cup of coffee, leaving the Royal McBee to chug along noisily. When Lorenz returned, he found that the computer program, rather than repeating its previous patterns, now was cranking out very different kinds of "weather." After checking out the obvious explanations, like a blown vacuum tube, the mathematician began looking more closely at the numbers. An extremely small but enormously significant difference emerged. As programmed, the computer was operating with the variables defined way out to six decimal points, although he had instructed it, when printing out its results, to round off the values to three decimal points, to thousandths. Lorenz had punched in these slightly abbreviated values as the new initial conditions. Now he realized suddenly that great differences in the weather patterns had grown from the minute differences in the two sets of figures.

"This was exciting," Lorenz recalled. "If the real atmosphere behaved in the same manner as the model, long-range weather prediction would be impossible, since most real weather elements are certainly not measured accurately to three decimal places. Over the following months, I became convinced that the lack of periodicity and the growth of the small differences were somehow related and I was eventually able to prove that, under fairly general conditions, either type of behavior implied the other. Phenomena that behave in this manner are now collectively referred to as chaos. This discovery was the most exciting event in my career."

Lorenz had discovered a characteristic that is inherent not only to the atmosphere but to many systems, although it would take years for physicists and biologists and others to locate in the *Journal of the Atmospheric Sciences* and to absorb the meaning for their own disciplines of his groundbreaking 1963 paper, "Deterministic Nonperiodic Flow." The key phrase is "sensitive dependence on initial conditions." Small differences in initial conditions lead to large differences in outcome. This principle would come to be known as the *butterfly effect,* a coinage inspired by an only slightly fanciful question Lorenz posed as the title of a talk to the American Association for the Advancement of Science in 1972: "Predictability: Does the Flap of a Butterfly's Wings in Brazil Set Off a Tornado in Texas?"

The principle of chaos ranges far and wide over the sciences. All of the signal noise that scientists had been smoothing out for the sake of finding patterns turned out to be more important than anyone supposed. Unstable systems of multiple variables act in a way that confounds intuition: small differences in input don't lead to small differences in output, but inevitably they lead to big ones. Writing in 1994, meteorologist Stanley David Gedzelman, at the City College of New York, described the impact of the discovery in the magazine *Weatherwise.* "For centuries scientists put on blinders, ignoring quirky and unpredictable phenomena in order to ferret out a tenuous sense of order in nature," he wrote. "This is the triumph of science, but it made us deny that chaos lurks everywhere. Ed Lorenz not only opened our eyes to the ever-present chaos in nature but also found its governing principles. He crowned 20th century meteorology with a discovery that irreversibly changed our view of the world."

Chaos eventually would become the business of physicians studying the irregular beats of the heart, of biologists trying to understand sudden spikes or collapses of animal populations, and of physicists, astronomers, chemists, and geologists reexamining unpredictable behavior in their fields.

To meteorology, its meaning was readily apparent, as Lorenz recognized immediately in 1961. He concluded in his landmark 1963 paper: "In view of the inevitable inaccuracy and incompleteness of weather observations, precise very-long-range forecasting would seem to be nonexistent." No matter how perfect the observations or how perfect the implementation of the laws governing the turbulent processes of the atmosphere, detailed weather forecasting beyond several days would still be impossible. The features of weather are variable in both space and time, differing from one place to another, one moment to the next.

Drawing the line between what weather forecasters can do and what they can't, finding the intersection of the upper limits of detailed daily forecasts and the lower limits of chaos, has been the subject of continuing research since the 1970s. Computer-generated Numerical Weather Prediction occupies a central role in all this. The computer giveth, and the computer taketh away. Before Numerical Weather Prediction, the effects of chaos were lost in the imprecision and subjectivity of forecasting. As meteorologist Philip D. Thompson later related, the issue of predictability "wasn't even a very sensible question" until numerical methods came along. As Lorenz and subsequent researchers illustrated, the main instrument in analyzing the problem was the same that revealed it: the Numerical Weather Prediction model.

While many meteorologists at the time were disinclined to pursue a "bad news" study of their practical limits, the meaning of Lorenz's work was immediately recognized by the great Jule Charney, his friend and fellow researcher down the hall at MIT. Lorenz recalled that in the mid-1960s Charney was actively promoting the big international meteorological study known as the Global Atmospheric Research Program. One of the big selling points for government financial support was the aim of GARP to produce two-week weather forecasts. The discovery of chaos posed an immediate threat to such a widely heralded ambition. Lorenz recalled that Charney "became concerned that two-week forecasting might be proven impossible even before the first two-week forecast could be produced, and he managed to replace the aim of making these forecasts with the more modest aim of determining whether such forecasts were feasible."

Charney discussed the possibility of chaotic behavior in 1964 at a conference in Boulder, Colorado, of leading meteorological researchers from 10 nations. Between sessions, he persuaded the global circulation modelers to test their computer models for sensitive dependence, as Lorenz had done, by running two forecasts with minor differences in initial conditions. The modelers reported back that on average, the small errors in temperature or wind pattern doubled in five days. "The five-day doubling time seemed to offer considerable promise for one-week forecasts, but very little hope for one-month forecasts, while two-week forecasts seemed to be near the borderline," Lorenz concluded.

The discovery that atmospheric motions are so sensitive to small differences in variables also caused the retirement of the so-called analog method as a tool for forecasting the details of daily weather. A mathematical study by Lorenz in 1969 proved the futility of looking for a set of past conditions that so closely matched the present that it told the future. "There are numerous mediocre analogues but no truly good ones," he concluded. The effect of small errors were difficult to calculate, he said, because truly small errors were hard to find. The recognition that small differences in initial conditions grow into large differences in weather cast a whole new light on the subject. A forecasting technique based on the apparent similarity of bygone conditions was not going to trick a chaotic atmosphere into revealing the weather of its future.

While the two-week range remains the most commonly accepted limit for detailed weather forecasting on theoretical grounds, modern forecasting's practical limit remains closer to a single week. Extended forecasts of increasing skill have been issued by the European Centre for Medium-Range Weather Forecasts in Reading, England, where the most powerful supercomputers are employed, and by the U.S. National Weather Service's National Centers for Environmental Prediction in Camp Springs, Maryland, which always seems to lag behind in computer horsepower and forecasting skill. In any case, a detailed two-week forecast currently is so eroded by multiplying error effects that under the best circumstances it offers only a modest advantage over pure guesswork.

Short-range accuracy has been boosted by numerical models of greater sophistication running on computers of greater power and speed, and over the first 40 years of operational Numerical Weather Prediction the range of accuracy has been extended from three days to about six. Well short of the two-week borderline suggested by studies of chaos, the signs of diminishing returns are already beginning to show. Extending the same level of detail and skill out beyond four or five days is coming at an increasingly high cost. And still there are surprises, embarrassing busted forecasts that every winter draw attention to the state of a science that modern users of its products rather remarkably take for granted. A man who before everyone else understood the meaning of chaos to the science of weather forecasting, Edward Lorenz has a different point of view on the subject: "To the often-heard question, 'Why can't we make better weather forecasts?' I have been tempted to reply, 'Well, why should we be able to make any forecasts at all?'"

27

Tetsuya Theodore Fujita

Divining the Downburst

•

EASTERN AIRLINES FLIGHT 66 from New Orleans was approaching John F. Kennedy International Airport on the afternoon of June 24, 1975. Rain was falling intermittently and there were thundershowers in the New York area, although winds at the airport measured only seven miles per hour as the Boeing 727 began its final approach toward Runway 22L at 4:01 P.M. Controlled by the instrument landing system, the airliner followed a routine glide path down to 400 feet when it hit a wall of heavy rain and a blast of wind. Its speed slowed drastically in just the time it takes to read these words, and just as quickly it sank. The crew fought to regain control, to bring the nose up and generate thrust, but it was too late. A wing clipped through one and then several of the stanchions supporting the airport approach lights. The plane crashed in a fireball, the fuselage eventually skidding across Rockaway Boulevard, half a mile short of the runway. The nation's worst airline disaster at the time killed all but 11 of the 124 passengers and crew members.

Everything about the disaster pointed to weather as its cause. But nearly a dozen aircraft had landed safely on Runway 22L within 30 minutes of the crash. Exactly what circumstances of weather had pushed the landing of Eastern Flight 66 beyond the performance limits of the Boeing 727 or its crew? Investigators for the National Transportation Safety Board cited "adverse winds" and, in the clarity of hindsight, a contributing cause: "the continued use of Runway 22L when it should have become evident to both air traffic control personnel and the flight crew that a severe weather hazard existed along the approach path."

Neither the federal investigators or anyone else at the time could resolve a central mystery of the weather that day. What was the real nature and the dimensions of the hazard? Although thunderstorms had been studied in considerable detail in the late 1940s, air travel had changed, and almost nothing in the meteorological literature could account for the effect of wind shear on the landing of a heavy jet. What weather conditions explained the perfectly safe landing of an airliner one moment and a disaster the next? Where did such fatal currents come from so suddenly and in such close proximity to an airport that was registering six-knot winds? These were the questions on the minds of flight safety experts at Eastern Airlines and elsewhere. They put them to the one man in the country they thought might be able to find an answer.

At the University of Chicago, meteorologist Ted Fujita was famous for his unique studies of the winds of thunderstorms. He would think of himself as someone who had walked through life stepping on stones marked "good luck," but Fujita had made his own luck. In Japan, he had purchased an English typewriter and finger-pecked a translation of his "micro-analytical study" of a thunderstorm. In 1950, he sent it off to Horace Byers, chairman of the Department of Meteorology at Chicago. Having just completed an extensive U.S. government–sponsored study of thunderstorms, Byers had been so impressed that he invited Fujita to join his faculty. For all his luck, Fujita worked harder and longer than anyone around him. His students in the Department of Geophysical Sciences sometimes called him "Superfuji," although not to his face.

His meticulous description of the 1957 tornado outbreak in Fargo, North Dakota, had become a classic in the new field of *mesoanalysis,* the study of storms below the continental scale of synoptic meteorology and beyond the resolution of numerical computer models. In 1971, with his wife, Sumiko, he had fashioned the Fujita scale, or F-Scale, for classifying the intensity of tornadoes according to their wind speed and the damage they do. Already it was gaining wide acceptance. To the public, he had become "Mr. Tornado."

At first, Ted Fujita had his doubts about wind shear as the explanation for the crash of Flight 66. But then he recalled a few striking images in photographs he had taken in April 1974 when he was investigating the worst tornado outbreak on record, the famous "super outbreak" of 147 tornadoes that had ravaged 11 southern and midwestern states. In some instances, uprooted trees had been felled in a small but distinct starburst pattern, not a twisted whirl but splayed out by a wind that must have shot straight down out of the clouds and burst across the ground.

Fujita had other experience with impacts of this configuration. As a professor in Japan, before his move to the United States in the 1950s, he had been called upon to survey the damage done by the atomic bomb that exploded over Nagasaki in 1945. From the large-scale starburst pattern of burn and shock wave damage, Fujita had surmised, and so informed Japanese authorities, that the devastation of Nagasaki was the result of the detonation of a single airborne device.

At the beginning of the Christmas recess in 1975, Fujita sat down and began analyzing the data from Flight 66, and before long, a hypothesis was taking shape. Fujita approached the problem as he did all of the mysteries of storms, working like a great detective, inspired by intuition. "There was an insight he had, this gut feeling," recalled James W. Wilson, a senior scientist at the National Center for Atmospheric Research in Boulder, Colorado, and an important Fujita collaborator. "He often had ideas way before the rest of us could even imagine them."

Fujita reviewed the radio messages that passed between the landing aircraft and the Kennedy tower, noting which pilots had reported encountering troublesome winds on Runway 22L and which had not. Here was a central mystery: could the scale of wind shear be so small and so brief and its effects so discrete? He analyzed weather conditions from every available angle: from satellite images, from radar scans, from conventional synoptic weather maps depicting simultaneous observations over large areas. From these various, individually sparse sources, he began synthesizing the data in his unique and creative way.

In March 1976, he published the results of his study, identifying the cause of the crash of Flight 66 as a previously unidentified feature of a thunderstorm he called a *downburst*. "Detailed examination of meteorological conditions revealed that the growth rate of the JFK Thunderstorm was at its peak when the accident occurred," Fujita wrote. "The radar echo of the storm appeared as a spearhead moving faster than any other echo in the vicinity. Hidden in the spearhead echo were four to five cells of intense downdrafts which are to be called 'downburst cells.' Apparently, those aircraft which flew through the cells encountered considerable difficulties in landing, while others landed between the cells without even noticing the danger areas on both sides of the approach path."

The paper set off a controversy among meteorologists. Many were skeptical, arguing that Fujita was merely giving a new name to the downdraft phenomenon that had been known for decades, or that he was mistaking the wind for the well-known gust front of squall line thunderstorms. Scientists who might have wanted to put the hypothesis to an independent test found that they didn't have enough information. Certain assumptions had to be taken on faith.

Fujita's unorthodox approach left him open to attack. He was working at the edge of scientific convention, rushing hypotheses into print without peer review or the volume of data that others would require. "He was a controversial character at times because of the way he did his science," Wilson recalled in an interview. "From limited observations he would theorize how things work, and it often was left to the rest of us to come along and prove his theories." Fujita's oft-stated dictum must have sounded cavalier to some: it was his job to get his ideas out, and it was up to others to prove him wrong. If he was right half the time, meteorology would be well served. For all the swagger, however, he was not an easy mixer in the give-and-take of Western science. The world was divided into bursters and anti-bursters. Perhaps because of his early training in the oriental formalisms of prewar Japan, public criticism was a personal matter. The scientist would recall that the controversy surrounding his downburst concept "led to several sleepless nights."

Unlike most scientific debates, however, the question of wind shear had a clear and urgent social imperative and a very interested public constituency. Some airline pilots hailed the paper as a breakthrough, offering their own testimony to having been victims of such winds while flying through even innocuous showers. At the nation's airports, meanwhile, wind shear was being blamed for the crash of a heavy jet every 18 months, on average. Between 1964 and 1983, it would cause eight airliner accidents and kill 514 passengers and crew members. Federal flight safety investigators called it the country's most dangerous threat to the safety of commercial aviation. On August 1, 1983, one of the strongest downbursts ever recorded struck Andrews Air Force Base in Maryland just six minutes after Air Force One, carrying President Ronald Reagan, had touched down.

In the mid-1970s, however, even those willing to accept the new discovery had to admit that the heralded downburst, or *microburst* as it came to be called, really had not been observed by anyone, including Ted Fujita. The crashes were terribly real, of course, and wind shear was undeniably a threat; but there were serious doubters among airline safety regulators and others about microbursts. If microbursts were anything more than figments of Superfuji's fertile imagination, somebody was going to have to prove it. Backed by Horace Byers, Fujita gained critical early support from key officials in Washington at the National Science Foundation and scientists in Boulder at the National Center for Atmospheric Research (NCAR), including radar expert Robert J. Serafin.

The first major break came on May 29, 1978, when Fujita and Jim Wilson observed their first microburst on a Doppler radarscope near

Yorkville, Illinois. Three Doppler radars had been deployed in a triangle near Chicago. Before long, they had the proof they needed: a total of 50 microbursts were detected by the network in the summer of 1978. Because they were thinking of microbursts as rare events, the scientists had made the legs of the triangle too large, the resolution of the network too coarse, for the kind of detail they sought. Still, even with this evidence, there were doubts about the threat posed by microbursts within the Federal Aviation Administration. The clincher came in the summer of 1982, when Fujita and the NCAR scientists experimented with another, more densely arrayed Doppler network around Denver's Stapleton International Airport, where the scientists were able to observe directly the effect of wind shear on aircraft. During this project, a total of 186 microbursts were observed, establishing the phenomena not only as real, but much more common than anyone had supposed.

As Serafin related, the Joint Airport Weather Studies Project at Stapleton led to a flurry of research that for the first time described the structure, evolution, and cause of microbursts. Developed through the mid-1980s, the findings made plain that decades of investigations into thunderstorms had failed to detect these transient events because they are too brief and too small for conventional monitoring systems. Moreover, as Byers had pointed out to Fujita, during the large and expensive Thunderstorm Project during the 1940s—which employed aircraft, balloons, and radar—nobody was looking for such things.

While the processes of cooling by evaporation and warming by condensation organize themselves into the updrafts and downdrafts of a maturing thunderstorm, a microburst is born of the runaway cooling of dry winds flowing into the midsection of a developing cloud. The dry air absorbs some of the cloud's moisture, quickly cooling and becoming more dense and sinking toward the ground, gaining momentum as it falls. "The outbursting of the air flow when a vertical column of downward moving air strikes the ground is very similar to what is observed when a water hose is pointed at a hard surface," Serafin noted in a recent description of microbursts. Wind hits the ground like a shock wave, bursting outward in all directions from its center.

Microbursts can generate wind speeds greater than 150 miles per hour and focus on an area as narrow as one-quarter mile in diameter, although more typically they are in the range of two and a half miles wide. They can occur anywhere in the country. In the eastern two-thirds of the United States, where high humidity is common during summer months, a contrasting dry flow aloft can generate rain-soaked microbursts that are imbedded in the downpours of thundershowers. In the arid West, dry microbursts can develop in "weakly convective" altocumulus clouds that never become storms and produce only light *virga*, rain that evaporates before reaching the ground. As Wilson related, scientists were stunned by Fujita's documentation of "intense microburst winds from innocuous high-based clouds over Colorado" during the Stapleton experiments.

Even when microbursts are detected, the speed of events and their transient nature raised serious questions in the 1980s about the ability of airport flight controllers and pilots to respond. Studies at NCAR by Wilson and John McCarthy found that from beginning to end, the process leading to a microburst typically runs its course in only 15 or 20 minutes, and that the wind shear hits its peak within 5 minutes of reaching the ground. "These numbers made it clear that very rapid means were required to detect microbursts and communicate warnings to pilots," Serafin observed.

Wet or dry, a greater, more insidious threat to a jetliner approaching or lifting off a runway would be hard to contrive. Only in the most technical sense does the aeronautical term *wind shear* describe what an airline pilot encounters. Three lethal contrary bursts hit within seconds. First a wall of headwind blasts against the plane and lifts the nose above the prescribed glide path. As the pilot dips the nose back down, the full force of the downburst pushes the plane toward the ground. Immediately thereafter, a powerful tailwind bucks the nose downward and further slows the plane. Dangerously low to the ground, and dangerously near the absolute airspeed limit, a pilot fights to bring lift and thrust to an aircraft that suddenly wants to pitch and roll in ways that have nothing to do with its forward motion or the controls being applied.

Equipped with new understanding of the threat, researchers in the mid-1980s began advocating improvements in airport and flight-training procedures and calling for new monitoring equipment to warn off pilots when microbursts are detected. Their case was made most forcefully and tragically at Dallas/Forth Worth International Airport the afternoon of August 2, 1985. Delta Airlines Flight 191, a Lockheed L-1011 from Ft. Lauderdale, Florida, crashed while landing in a microburst, killing 137 of its 167 passengers and crew members.

The Federal Aviation Administration (FAA) moved with unusual dispatch to transfer the basic science to the aviation industry, developing new microburst detectors and pilot-training programs. Work by scientists and engineers in the Lincoln Laboratory of the Massachusetts Institute of Technology led to the development of an automated expert system known as the Terminal Doppler Weather Radar System, which has been installed in most major airports around the United States and elsewhere around the world. It is in the nature of these events that they can be predicted only a few minutes in advance, but a few minutes is enough time to divert an airliner out of harm's way. The new airport radars provide what Wilson called "timely and unambiguous warnings of hazardous wind shears and microbursts." At the same time, scientists quickly organized special, intense new FAA-mandated training programs that taught commercial airline pilots how to recognize the hazards of microbursts, and how to survive them.

The change was soon apparent at the nation's airports, where a dramatic, lifesaving turnaround was accomplished. Before 1985, microbursts were being blamed for an airline disaster every 18 months. After 1985, the next microburst-related accident was not until 1994. Just 10 years after the first discovery of the microburst by Ted Fujita in 1976, practical lifesaving solutions were in place. Observing the pace of events, Wilson said, "There's just never been anything to parallel that in meteorology before."

As was typical of his style and pace, once he was satisfied that his hypothesis was sound, Fujita left much of the work of developing the full scientific case for microbursts to others, especially Wilson and the NCAR scientists, and Roger M. Wakimoto, his University of Chicago student. Mr. Tornado had other mysteries of the wind to investigate. But his role in a life-saving discovery was the kind of event that makes scientists' dreams. And his controversial, intuitive style of science probably led to the discovery sooner rather than later.

"Fujita is the person who was responsible for first proposing and eventually proving the existence of the downburst," Wilson wrote. "There is little doubt that the downburst would have eventually been discovered without his contribution, but at what expense?"

Mr. Tornado suffered from diabetes and often endured severe pains in his legs, although he was never to complain about it, and almost no one outside his family was aware of his chronic condition. He retired from teaching at the University of Chicago at the age of 70 but continued his research. Even as illness confined him to his bed, he continued to work with the aide of research assistants, Superfuji to the end. He died at his home in Chicago on November 19, 1998, at the age of 78.

28

Ants Leetmaa

Out on a Limb

•

IT WAS 10 A.M. on June 17, 1997, and for the director of the national Climate Prediction Center, a defining moment had arrived. The bright lights came on, and he walked to the podium of the Murrow Room at the National Press Club in Washington. As he arranged the notes for his presentation, he could feel the heat of the lamps already penetrating his fair scalp.

The scientist bureaucrat, an oceanographer by training, was taking a big risk in a town and in a business that don't like even little risks. He was going to make a prediction that had not been made before—not accurately—about the weather that Americans could expect far into the future. And no matter how carefully he phrased his words, he knew how easily he could be made to sound silly or outlandish. As well as any climate scientist, he knew the American media's long tradition of lightheadedness when it comes to reporting the weather.

By conventional government standards, his timing was all wrong, because even if everything he said at the day's news conference turned out to be true, to be accurate, it would be several months before anybody could be sure. Along the way, meanwhile, many of his scientific colleagues would be interviewed and would not support his conclusions. And if he turned out to be wrong, well, Ants Leetmaa would not be the first scientist to be chewed up and spit out by the U.S. government.

In any case, it wasn't going to be easy to explain. Like some conjurer, he was about to stitch some time and space together in a way that most people were not accustomed to doing. On the basis of information about ocean surface temperatures thousands of miles away in the tropical Pacific during the month of May, he was going to stand before national television cameras and purport to tell the country how the weather was going to go across the United States the following winter. He would talk about the kinds of temperatures and precipitation to expect, and warn citizens in the western and southern regions of the country that things could get bad.

Ants Leetmaa had invited his boss and others at the National Oceanic and Atmospheric Administration (NOAA) to join him at the podium that day. This offer to share the limelight was the right thing, of course, the politic thing, really an essential part of getting along in the federal bureaucracy. But there were no takers for this auspicious occasion. On June 17, 1997, the first day of winter was 27 weeks away. No, thank you very much. The powers that be at NOAA were going to let the long-rangers, the renegade El Niño people, have this news conference all to themselves.

Information Officer Stephanie Kenitzer counted cameras and heads in the audience and knew there would be wide coverage of Leetmaa's announcement. An Associated Press (AP) rewrite of the basic press release she had issued was already on the wire.

"Strong El Niño conditions are currently developing in the tropical Pacific," the wire service report said. "The warm event will bring wetter, cooler weather for the southern half of the United States from November through March, while the northern part of the country from Washington east to the western Great Lakes will experience warmer than normal temperatures, according to predictions from the National Oceanic and Atmospheric Administration's Climate Prediction Center."

Closely following the wording of the press release into an unfamiliar subject, the AP story described El Niño as "an abnormal state of the ocean-atmosphere system in the tropical Pacific having important consequences for weather around the globe. Among these consequences are increased rainfall across the southern tier of the United States and in Peru, sometimes resulting in destructive flooding; and drought in northeast Brazil, southeastern Africa, and the west Pacific."

In June 1997, Leetmaa had a Washington press corps and a whole country to educate. "These are the typical weather impacts during a warm phase based on an average of all El Niño events," he said. "This El Niño event is shaping up to be similar to the strong events of 1957, 1972, and 1982–83. During those years, many sections of the southern half of the United States, including California, experienced above normal rainfall from September through the following May."

Leetmaa was inclined to couch his remarks in the jargon of climate prediction. He could comfortably talk about sea-surface temperature anomalies—about coastal upwelling, anomalous winds, intraseasonal oscillations and Kelvin waves, outgoing long-wave radiation, and geopotential heights. Professional meteorologists in the Washington audience might have preferred such a presentation. But the correspondents were not in the national capital to cover climate science and would not have understood him. And what was the point of making so public a prediction about the coming winter so early in the year unless he was going to be understood and was willing to attract attention? He was going to have to shed the jargon and speak plainly.

"We're pretty concerned," he said.

"This is not the average El Niño."

"It's the first time it's reached this magnitude this early in the summer season."

"This is looking much more like a big one than a small one."

"This is shaping up to have a large, significant impact all across the globe."

"California is going to be wet, the Southwest is going to be wet, the Southeast is going to be wet."

James Gerstenzang of the *Los Angeles Times* asked about El Niño's impact on the hot Santa Ana winds that often bring fall brush fires down on southern California. Leetmaa said it would curb that pattern of winds, but "it'll be so wet they won't matter."

The plain talk came at a price. Many climate scientists were already skeptical of the claims about El Niño and its impact on world weather, and they were willing to say that Leetmaa had gone too far, that he had over-stated the link between ocean temperatures now and weather in the future. Gerstenzang observed in his story a widely held view of oceanographers and meteorologists: " . . . looking at current conditions and leaping to predictions claiming anything more certain than a likelihood is a particularly uncertain course."

Chris Cappella at *USA Today* asked William Gray of Colorado State University about Leetmaa's suggestion that the El Niño could dampen 1997 hurricane activity in the Atlantic. The guru of hurricane season forecasting replied, "I do not think El Niño will play a dominant role in reducing storm activity this year." (As it happened, Gray would produce one of his worst seasonal forecasts that year.)

"We had a lot of disbelievers," Kenitzer recalled. In the audience at the news conference was Richard Hallgren, executive director of the American Meteorological Society, who earlier had publicly rebuked the

El Niño scientists for their claims. "The El Niño guys were kind of like renegades, I mean, they're radicals," Leetmaa would say later, and he would remember the sight of Hallgren in front of him: "He shook his head disbelievingly at the statements that were coming out of my mouth."

Finally the lights were out and it was over. Leetmaa walked over to Kenitzer and muttered, "God, I went out on a limb here."

He had gone against the conventional wisdom of meteorology, against the advice of some colleagues, against the conservative traditions of science and government. Why did he do it? Ants Leetmaa saw the risk as part of his job. For the first time, scientists had detected conditions in the tropical ocean that would change the face of the coming winter far in advance of their potentially deadly effects around the world. Not just a little, in this case, but a lot. Lives were at risk. There was property to protect. As far as Leetmaa was concerned, the Climate Prediction Center had a fairly clear-cut mission under the circumstances, and so did the new science.

In the months ahead, Leetmaa would grow fond of a book by the economist Peter L. Bernstein, *Against the Gods: The Remarkable Story of Risk,* and in presentations he would quote from its introduction: "The revolutionary idea that defines the boundary between modern times and the past is the mastery of risk: the notion that the future is more than a whim of the gods and that men and women are not passive before nature."

"We used to think of these things as random events or acts of God," Leetmaa would say. The new science is not really about this El Niño, or any El Niño. It is about the boundary between modern times and the past. The science of seasonal forecasting, of climate prediction, is about nothing more or less than seeing farther into the future than ever before. It is about bringing a powerful new vision to life—the ability to know about next winter. It is about mastering risk, about not being "passive before nature."

Like a riverboat gambler, Leetmaa had just put the credibility of this infant science on the line in a very public way. He was at the table of a crooked crap game with the dice in his hand. Sure, he knew the dice were loaded, and like any good long-ranger he knew that over time the skewed probabilities would play themselves out. Leetmaa's problem was that he had only one roll. In the coming months, seasonal forecasting could be heralded as a reality, or dismissed as a distant dream.

It was what Leetmaa would call an "opportunity forecast," a prediction that could be made only because certain inherently predictable conditions were in place. The science of climate prediction, or "seasonal foreshadowing," first suggested by the statistical research of Sir Gilbert

Walker early in the century was coming to fruition. The line of thinking from Walker to Jacob Bjerknes to Jerome Namias about how the ocean and the atmosphere are so closely related to each other had led to this: conditions of temperature and pressure in the equatorial Pacific, when they reach significant extremes above or below normal, lead almost invariably to a certain set of anomalous seasonal weather conditions around the globe several months hence.

More opportunity forecasts are likely to present themselves as scientists refine their understanding of other cyclical large-scale ocean-atmosphere anomalies. These are the stuff of the new science of climate prediction: changes in planetary characteristics with consequences that move from one timescale to another. On the scale of weeks, an atmospheric wave that circles the equator every 45 to 60 days provokes tropical thunderstorm activity that leads to winter storminess over the Pacific Northwest. On the scale of months, a shifting intensity in stratospheric winds, defined by John M. Wallace at the University of Washington as the Arctic Oscillation, moves tracks of storms between different latitudes of the Northern Hemisphere. On the scale of years, an ocean temperature pattern defined by scientists at Scripps Institution of Oceanography as the Pacific Decadal Oscillation shifts the intensity of winter seasons over the western United States. And every 3 to 7 years, on average, there is the El Niño/Southern Oscillation and the global reach of its seasonal weather effects.

In the early summer of 1997, Ants Leetmaa looked at Pacific Ocean sea-surface temperatures that were rising faster than anyone had ever seen and felt compelled to issue an alert to public disaster agencies across the country. Still, this was not the way the director of the Climate Prediction Center or anyone else in climate science or in NOAA would have planned to roll out a new forecasting product. The equatorial Pacific that drives so much of the world's weather had been a poorly defined, unstable mess through much of the 1990s, an inconclusive muddle of shifting temperature patterns that had been fooling computer modelers and theorists. It had everyone looking fairly clueless for a few years. Computer climate modelers, in fact, had not seen this big event coming. In the spring of 1997, however, the planet was forcing the issue.

Leetmaa was relying heavily on two recent developments. Over a 10-year period ending in 1995, researchers taking part in a project known as the Tropical Ocean Global Atmosphere program had built up a much clearer picture of the geometry of the Pacific Ocean. When sea-surface temperatures change along the equator, the atmosphere quickly responds, and now they could see how the atmosphere flings these effects out as weather events around the world. This big international research campaign had left in place a powerful detection system, a network of instrumented buoys that is now spread across the band of ocean that girdles a third of the globe. The equatorial Pacific Ocean is wired like a heart patient. In the spring of 1997, the instruments of the new network were sending an unmistakable signal: something like a massive coronary was under way.

Since December 1996, three big episodes of powerful wind bursts in the western Pacific had reversed the trade winds and sent deep, subsurface Kelvin waves rippling through the ocean, changing its thermal structure as they rolled toward South America. A great reservoir of warming was set in motion, traveling at about walking pace. In March, April, and May, as the waves arrived, sea-surface temperatures in the eastern Pacific shot up as if the ocean were catching fire. Scientists had never seen anything like it. Never had they seen ocean surface temperatures change so quickly or reach such warm levels so early in the year. By May, sooner than ever, Leetmaa knew he had a big El Niño on his hands.

At the heart of Leetmaa's gamble was a proposition that computer modeling strongly suggested but the planet had not as yet confirmed: one powerful El Niño acts pretty much like another, at least in the sense that its major impacts are spread similarly around the globe. If he were right, developing conditions would play out much like they did during the 1982–1983 monster, "the El Niño of the century." If he were wrong, he could be very wrong, because while the pattern of impacts might be different, he had no doubt about their impending power. Some people were in for a hell of a winter. Exactly who, exactly where, was the problem.

Across the country, as the news spread, local National Weather Service meteorologists found themselves being asked questions that they were not prepared to answer. How was El Niño going to affect Sacramento and Buffalo and Peoria? Caught off-guard, many of these trained weather scientists naturally expressed skepticism. Sure the atmosphere responds to *boundary conditions,* the temperatures of the ocean or the ice or the terrain it flows over. In conventional meteorology, in daily forecasting, however, the storm is the thing, and the storm is not an oceanic event.

Ants Leetmaa trained in oceanography at the Massachusetts Institute of Technology, the greatest meteorological science university in the country, under the tutelage of Henry Stommel, a leading theoretical oceanographer. MIT is where Jule Charney laid the theoretical groundwork for Numerical Weather Prediction—computerized forecasting—and where Norman Phillips wrote the book on the subject. Charney, Stommel, Phillips: these great scientists were not trained in the idea that the tropical ocean played such a decisive role in world weather. "None of those three guys understood that you could make a forecast of El Niño," Leetmaa would say. And none of the nation's operational meteorologists were schooled in the thought that the ocean and the atmosphere are so closely coupled to one another. "This is a very new thing."

In June, there was no way for Leetmaa or anyone to know how quickly and how thoroughly scientists would lose control of the message they were trying to convey. El Niño took on a life of its own in the popular culture. Before long, climate science and the rhythms of the planet had very little to do with what was going on. As time passed, temperatures in the tropical Pacific continued to rise, and the popular media became more fascinated with the subject. Leetmaa found himself sharing the television cameras with bureaucrats and then politicians. For a while, in early fall, events seemed to be driven not so much by the impending winter as by the impending international meeting on global warming in Japan in December and the 1998 elections in California. For a time, the climate scientists were nearly crowded off the podium. Widely publicized government summits on the subject of El Niño preparedness in California took on inevitably partisan political slants. With the upcoming Tokyo meeting in mind, Vice President Al Gore raised the political ante, suggesting the coming El Niño was going to be a taste of the kind of winters the nation can expect from a world of unrestrained greenhouse gas emissions. With this suggestion, political commentators now knew what to make of El Niño.

There were silly debates—loud, partisan, and beside the point—reinforcing old feelings among veterans of such wars that the mass media is no place to discuss science. It was going to be the "Climate Event of the Century" and a harbinger of global warming, or it was all just a bunch of liberal, big-government baloney. Across the country, El Niño was more a theme of television and radio programming than it was a subject of climate science, and the new knowledge it represented about how the world's weather works took a real mauling. Sadly, the real El Niño events of the time—the fires in Indonesia and the killing drought and famine in Papua New Guinea—went largely unnoticed in the U.S. media din.

A few scientists tried to bring the subject back to its roots. One was Edward S. Sarachik, a noted atmospheric scientist at the University of Washington. "I don't want you to go away thinking the story is just that this is the biggest event of the century," Sarachik told a news briefing in Seattle. "It is, but we've also learned how to use climate information, how to produce climate information. And I think that has the capability of altering our lives more than a single El Niño."

Across the country, climatologist Jagadish Shukla at George Mason

University, who had been a graduate student of Jule Charney, was trying to make the same point: "What has been lost in this whole media hype is, we are witnessing the birth of a new science. And that new science is climate prediction."

And then the rains came, of course, and winter storms battered California and the Southwest and the Southeast just like Leetmaa said they would. And the northern tier of the country experienced an unusually dry and warm winter, just as the Climate Prediction Center said it would in June of the preceding year.

Whether anybody remembered after all the hype and incessant media banter on the subject, it turned out to be a remarkably accurate forecast for the winter of 1997–1998 in the United States, the most accurate long-range forecast the government had ever issued. When it was over, the renegade El Niño scientists, the seasonal forecasters, were speaking to large and attentive audiences of their professional colleagues.

"I think there's been probably a revolution in terms of the local forecasters and the meteorological community," Leetmaa would say after one such session at an American Geophysical Union meeting in San Francisco, but he knew the ephemeral nature of such credibility, even among fellow scientists. "Now if we'd gone bust on the forecast, then they would have said, 'Well, we told you so. Those El Niño guys, they don't know what the hell they're talking about."'

Bibliography

•

Abbe, Cleveland. "Benjamin Franklin as Meteorologist." *Bulletin, American Philosophical Society*, April 20, 1906, pp. 117–128.

Abbe, Truman, *Professor Abbe . . . and the Isobars*. New York: Vantage Press, 1955.

Anderson, Katharine Mary. "Practical Science: Meteorology and the Forecasting Controversy in Mid-Victorian Britain." Ph.D. diss., Northwestern University, 1994.

Aspray, William. *John von Neumann and the Origins of Modern Computing*. Cambridge, Mass.: MIT Press, 1990.

Basu, Janet Else. "Jerome Namias: Pioneering the Science of Forecasting." *Weatherwise* (August 1984).

Bergeron, Tor. "Synoptic Meteorology: An Historical Review." *Pure and Applied Geophysics* 119 (1981), pp. 443–473.

Blench, Brian J. R. "Luke Howard and His Contribution to Meteorology." *Weather* 18 (1963), pp. 83–92.

Bolin, Bert. "Carl-Gustaf Rossby: The Stockholm Period, 1947–1957." Tellus 51 A–B (1999), pp. 4–12.

———, ed. *The Atmosphere and the Sea in Motion: Scientific Contributions to the Rossby Memorial Volume*. New York: Rockefeller Institute Press, 1959.

Bundgaard, Robert C. "Sverre Petterssen, Weather Forecaster." *Bulletin, American Meteorological Society* 60, no. 3 (March 1979).

Burton, Jim. "Robert FitzRoy and the Early History of the Meteorological Office." *British Journal of the History of Science* 19 (1986), pp. 147–176.

Cayan, Daniel R. "Tribute to Jerome Namias: The Scripps Era." *Bulletin, American Meteorological Society* 79, no. 6 (June 1998).

Cline, Isaac Monroe. *Storms, Floods and Sunshine*. Gretna, La.: Pelican Publishing, 2000.

Cohen, Bernard I. *Benjamin Franklin's Science*. Cambridge, Mass.: Harvard University Press, 1990.

Cressman, George. Interview, American Meteorological Society Tape Recorded Interview Project, 1992.

Davis, John L. "Weather Forecasting and the Development of Meteorological Theory at the Paris Observatory, 1853–1878." *Annals of Science* 41 (1984), pp. 359–382.

Davis, William Morris. "Some American Contributions to Meteorology." *Journal of the Franklin Institute* 127 (1889), pp. 104–115, 176–191.

Eliassen, Arnt. "Vilhelm Bjerknes and His Students." *Annual Review of Fluid Mechanics* 14 (1982), pp. 1–11.

Espy, James P. *The Philosophy of Storms*. Boston: Charles C. Little and James Brown, 1841.

Finley, John P. "Tornadoes." *Insurance Monitor* (1887), available at http://www.lib.noaa.gov/edocs/tornado/tornado.html.

Fleming, James Rodger. *Meteorology in America, 1800–1870*. Baltimore: Johns Hopkins University Press, 1990.

———, ed. *Historical Essays on Meteorology, 1919–1995*. Boston: American Meteorological Society, 1996.

———, ed. *Weathering the Storm: Sverre Petterssen, the D-Day Forecast, and the Rise of Modern Meteorology*. Boston: American Meteorological Society, 2001.

Friedman, Robert Marc. *Appropriating the Weather: Vilhelm Bjerknes and the Construction of a Modern Meteorology*. Ithaca, N.Y.: Cornell University Press, 1989.

Frisinger, H. Howard. *The History of Meteorology: To 1800*. Boston: American Meteorological Society, 1977.

Fuller, John F. *Thor's Legions: Weather Support to the U.S. Air Force and Army, 1937–1987*. American Meteorological Society, 1990.

Galway, Joseph G. "Early Severe Thunderstorm Forecasting and Research by the United States Weather Bureau." *Weather and Forecasting* 7, no. 4 (1992), pp. 564–587.

———. "J. P. Finley: The First Severe Storms Forecaster." Pt. I and II. *Bulletin, American Meteorological Society* 66, nos. 11 and 12 (1985).

Gedzelman, Stanley David. "Automating the atmosphere." *Weatherwise* (June–July, 1995).

———. "Cloud Classification before Luke Howard," *Bulletin, American Meteorological Society,* April 1989, pp. 381–395.

———. "Chaos Rules: Edward Lorenz Capped a Century of Progress in Forecasting by Explaining Unpredictability," *Weatherwise*, August-September, 1994.

Glaisher, James. *Travels in the Air*. Philadelphia: Lippincott, 1871.

Gleick, James. *Chaos: Making a New Science*. New York: Penguin, 1987.

Hasdorff, James C. "Interview of Lt. Gen. Donald N. Yates." United States Air Force Oral History Program, 1980.

Heninger, S. K., Jr. *A Handbook of Renaissance Meteorology*. Durham, N.C.: Duke University Press, 1960.

Howard, Luke. "On the Modification of Clouds." *Philosophical Magazine,* nos. 16, 17 (1803).

Hughes, Patrick. "FitzRoy the Forecaster: Prophet without Honor." *Weatherwise* (August 1988), pp. 200–204.

———. "The New Meteorology." *Weatherwise* (June–July 1995), pp. 26–35.

Hunt, J. C. R. "Lewis Fry Richardson and His Contributions to Mathematics, Meteorology, and Models of Conflict." *Annual Review of Fluid Mechanics* 30 (1998), pp. 13–36.

Hunt, Julian L. "James Glaisher FRS (1809–1903) Astronomer, Meteorologist and Pioneer of Weather Forecasting: 'A Venturesome Victorian,'" Quarterly Journal Royal Astronomical Society (1996).

Jahns, Patricia. *Matthew Fontaine Maury and Joseph Henry, Scientists of the Civil War*. New York: Hastings House, 1961.

Jewell, Ralph. "Tor Bergeron's First Year in the Bergen School: Towards an Historical Appreciation." *Pure and Applied Geophysics* 119 (1981), pp. 474–490.

Kutzbach, Gisela. *The Thermal Theory of Cyclones: A History of Meteorological Thought in the Nineteenth Century.* Boston: American Meteorological Society, 1979.

Lewis, John M. "Cal Tech's Program in Meteorology: 1933–1948." *Bulletin, American Meteorological Society* 75, no. 1 (January 1994).

———. "Carl-Gustaf Rossby: A Study in Mentorship." *Bulletin, American Meteorological Society* 73, no. 9 (September 1992), pp. 1425–1438.

———. "LeRoy Meisinger, Part I." With Charles B. Moore. "LeRoy Meisinger, Part II." *Bulletin, American Meteorological Society* 76, nos. 1, pp. 33–45, and 2, pp. 213–226 (1995).

Liljequist, G. H. "Tor Bergeron: A Biography." *Pure and Applied Geophysics* 119 (1981), pp. 409–419.

Lindzen, Richard S., Edward N. Lorenz, and George W. Platzman, eds. *The Atmosphere—A Challenge: The Science of Jule Gregory Charney.* Boston: American Meteorological Society, 1990.

Lorenz, Edward N., *The Essence of Chaos*. Seattle: University of Washington Press, 1993.

Ludlum, David. "The Espy-Redfield Dispute." *Weatherwise* (December 1969).

Maury, Matthew F. *The Physical Geography of the Sea*. Cambridge, Mass.: Belknap Press, 1963.

Mellersh, H. E. L. *FitzRoy of the Beagle*. London: Rupert Hart-Davis, 1968.

Middleton, W. E. Knowles. "P. H. Maille, a Forgotten Pioneer in Meteorology." *Isis* 56, 3, no. 185 (1965), pp. 320–326.

Namias, Jerome. Autobiography, in Namias Symposium. Scripps Institution of Oceanography, 1986.

———. "Long Range Weather Forecasting—History, Current Status and Outlook." *Bulletin, American Meteorological Society* 49, no. 5 (May 1968).

———. "The Early Influence of the Bergen School on Synoptic Meteorology in the United States." *Pure and Applied Geophysics* 119 (1981), pp. 491–500.

———. "The History of Polar Front and Air Mass Concepts in the United States—An Eyewitness Account." *Bulletin, American Meteorological Society* 64, no. 7 (July 1983).

Nebeker, Frederik. *Calculating the Weather: Meteorology in the 20th Century*. San Diego: Academic Press, 1995.

Phillips, Norman A. "Carl-Gustaf Rossby: His Times, Personality, and Actions." *Bulletin, American Meteorological Society* 79, no. 6 (June 1998), pp. 1097–1112.

———. "Jule Charney's Influence on Meteorology." *Bulletin, American Meteorological Society* 63, no. 5 (May 1982).

———. "Jule Gregory Charney." Biographical Memoirs. National Academy of Sciences, 1981.

Platzman, George W. "The ENIAC Computations of 1950—Gateway to Numerical Weather Prediction." *Bulletin, American Meteorological Society* 60, no. 4 (April 1979).

———. "A Retrospective View of Richardson's Book on Weather Prediction." *Bulletin, American Meteorological Society* 48 no. 8 (August 1967), pp. 514–550.

Pomerantz, Martin A. "Benjamin Franklin—the Compleat Geophysicist." *EOS, Transactions of the American Geophysical Union* 57 (1976), pp. 492–507.

Rasmusson, Eugene M. "Tribute to Jerome Namias: The Pioneering Years." *Bulletin, American Meteorological Society* 79, no. 6 (June 1998).

Richardson, Lewis F. *Weather Prediction by Numerical Process*. Cambridge: Cambridge University Press, 1922.

Roads, John O. "Jerome Namias." Biographical Memoirs. National Academy of Sciences, 1997.

Seeger, Raymond J. *Benjamin Franklin: New World Physicist*. New York: Pergamon Press, 1973.

Shaw, Sir Napier. *Manual of Meteorology*. Vol. I. Cambridge: Cambridge University Press, 1926.

Stagg, James Martin. *Forecast for Overlord, June 6, 1944*. New York: W. W. Norton, 1972.

Stommel, Henry. *The Gulf Stream: A Physical and Dynamical Description*. Berkeley: University of California Press, 1965.

Taba, Hessam. *The Bulletin Interviews*. Vols. I and II. Geneva: World Meteorological Organization, 1988, 1997.

Thompson, Philip Duncan. "A History of Numerical Weather Prediction in the United States." *Bulletin, American Meteorological Society* 64, no. 7 (July 1983).

Whitnah, Donald R. *A History of the United States Weather Bureau*. Champaign: University of Illinois Press, 1961.

Wurtele, Morton G. *Selected Papers of Jacob Aall Bonnevie Bjerknes*. North Hollywood: Western Periodicals, 1975.

Index

•